I0467872

Land-Use Portfolio Modeler, Version 1.0—Software Documentation and Tutorial

By Richard Taketa, Peter Ng, and Makiko Hong

Techniques and Methods Series 11 C4

U.S. Department of the Interior
U.S. Geological Survey

U.S. Department of the Interior
KEN SALAZAR, Secretary

U.S. Geological Survey
Marcia K. McNutt, Director

U.S. Geological Survey, Reston, Virginia: 2010

This report and any updates to it are available online at:
http://pubs.usgs.gov/tm/tm11c4/

For more information on the USGS—the Federal source for science about the Earth, its natural and living resources, natural hazards, and the environment, visit http://www.usgs.gov or call 1-888-ASK-USGS

For an overview of USGS information products, including maps, imagery, and publications, visit http://www.usgs.gov/pubprod

To order this and other USGS information products, visit http://store.usgs.gov

Suggested citation:
Taketa, R., Ng, P., and Hong, M., 2010, Land-Use Portfolio Modeler, Version 1.0: U.S. Geological Survey Techniques and Methods 11–C4, 54 p.

Contents

Figures

Land-Use Portfolio Modeler, Version 1.0— Software Documentation and Tutorial

By Richard Taketa[1], Peter Ng[2], and Makiko Hong[2]

Introduction

Natural hazards pose significant threats to the public safety and economic health of many communities throughout the world. Community leaders and decision-makers continually face the challenges of planning and allocating limited resources to invest in protecting their communities against catastrophic losses from natural-hazard events. Public efforts to assess community vulnerability and encourage loss-reduction measures through mitigation often focused on either aggregating site-specific estimates or adopting standards based upon broad assumptions about regional risks. The site-specific method usually provided the most accurate estimates, but was prohibitively expensive, whereas regional risk assessments were often too general to be of practical use. Policy makers lacked a systematic and quantitative method for conducting a regional-scale risk assessment of natural hazards. In response, Bernknopf and others (2001) developed the portfolio model, an intermediate-scale approach to assessing natural-hazard risks and mitigation policy alternatives.

The basis for the portfolio-model approach was inspired by financial portfolio theory, which prescribes a method of optimizing return on investment while reducing risk by diversifying investments in different security types. In this context, a security type represents a unique combination of features and hazard-risk level, while financial return is defined as the reduction in losses resulting from an investment in mitigation of chosen securities. Features are selected for mitigation and are modeled like investment portfolios. Earth-science and economic data for the features are combined and processed in order to analyze each of the portfolios, which are then used to evaluate the benefits of mitigating the risk in selected locations. Ultimately, the decision maker seeks to choose a portfolio representing a mitigation policy that maximizes the expected return-on-investment, while minimizing the uncertainty associated with that return-on-investment.

The portfolio model, now known as the Land-Use Portfolio Model (LUPM), provided the framework for the

development of the Land-Use Portfolio Modeler, Version 1.0 software (LUPM v1.0). The software provides a geographic information system (GIS)-based modeling tool for evaluating alternative risk-reduction mitigation strategies for specific natural-hazard events. The modeler uses information about a specific natural-hazard event and the features exposed to that event within the targeted study region to derive a measure of a given mitigation strategy's effectiveness. Harnessing the spatial capabilities of a GIS enables the tool to provide a rich, interactive mapping environment in which users can create, analyze, visualize, and compare different natural-hazard mitigation scenarios.

Background

The portfolio model was first demonstrated in a case study of an earthquake-induced lateral-spread ground failure in the Watsonville, California, area (Bernknopf and others, 2001). The study involved a probabilistic earthquake event similar to the magnitude 6.9 Loma Prieta Earthquake of 1989. The portfolio model was used to evaluate two mitigation policies—one that prioritized mitigation by land-use type and the other by hazard zone. The results of these policies were compared against the expected outcome from not performing any mitigation. The portfolio representing the hazard-zone rule yielded a higher expected return than the land-use rule, but it also experienced a higher standard deviation (measure of uncertainty); therefore, neither policy demonstrated a clear advantage. Nonetheless, the policies reduced expected losses and increased overall expected community wealth when compared to the existing policy of no mitigation.

The model has been applied in other benchmark studies, including a multi-hazard study involving earthquakes and floods in the District of Squamish, British Columbia, Canada (Wein, 2005) and another study involving earthquake-triggered landslides in Ventura County, California (Dinitz, 2008). These benchmark studies provided valuable insight and feedback used to help refine the model. The LUPM is still in the research stage, and, hence, it has and will continue to evolve, incorporating improved or advanced mathematical calculations, enhancing analytical techniques, and adapting functionality to meet newer research application requirements.

---

[1] San Jose State University Department of Geography and USGS Western Geographic Science Center.

[2] USGS Western Geographic Science Center.

Software Methodology

The LUPM v1.0 software estimates losses avoided and changes to community wealth based on specific hazard-mitigation strategies by assessing 1) the probability that a hazard event may occur, 2) the probability that the occurrence of that event will have an impact on features in the built environment (for example, damage from an earthquake), 3) the value of the features, and 4) the cost of protecting features from impact of the hazard event. The model calculates the losses avoided by applying the mitigation remedy to specified features, as well as the overall impact on community wealth. The model also calculates the variability of the results, based on the variances associated with the event and damage probabilities.

Policy choices primarily involve selecting locations to mitigate. These choices then represent decisions on where to invest the community's mitigation budget. Selection may be based on a variety of criteria, including feature attributes (such as potential for loss, value, likelihood of damage), feature locations (existing zones or regions, proximity to existing features, or location in or relative to ad hoc features), and/ or a combination of both attribute and location criteria. The results of the analysis provide an indication of losses that may be avoided through mitigation and can serve as a basis for comparing different mitigation selections.

The software can use data on partial losses for both mitigated and unmitigated features, based on calculations developed by Champion (2005 and 2008). If partial loss data are not provided, the software will assume 100 percent protection for mitigated features and 0 percent protection (complete loss) for unmitigated features impacted by the event. Partial losses enable the software to allow for less than 100 percent effectiveness of mitigation, as well as less than 100 percent damage, even with no mitigation, in determining the losses avoided.

Results from LUPM runs are reported in tabular form. The LUPM v1.0 software also provides procedures for saving different runs (that is, scenarios) and exporting results. The mechanism for saving and/or exporting results varies depending on the specific LUPM v1.0 package.

The LUPM v1.0 software can be adapted to the research problem being addressed, such as evaluating alternative flood-, landslide-, or earthquake-mitigation strategies. However, the basic process is the same for all applications. The operations include assembling available data, possibly reformatting or calculating values needed to run the portfolio analysis, formulating mitigation scenarios and running them through the LUPM v1.0 software, identifying appropriate ways to display and report the results, and developing an understanding of the implications of the results in a broader social, economic, and political context.

LUPM v1.0 is based on ArcGIS 9.2. Users should be familiar with ArcGIS operations, including using the feature-selection tools, managing geographic and tabular data, and creating map displays. Users who intend to use the LUPM Geoprocessing Tools package should also be familiar with the ArcGIS ModelBuilder environment. Experience using the Python scripting language is also helpful when using the LUPM v1.0 package.

Intended Audience

The LUPM v1.0 project is aimed at U.S. Geological Survey (USGS) research scientists and collaborators. It is not intended for a general audience, largely due to the exploratory nature of the modeler. We are just beginning to understand how it may be applied and how it can be connected to a larger world.

The research orientation for this version has a number of implications. First, the land use portfolio calculations are evolving. The model is being expanded to incorporate multiple hazards in a single analysis, to utilize spatial autocorrelation, to handle multiple time periods, and to handle regionally aggregated data. Second, the way in which a tool such as the LUPM may be applied is still evolving; this implementation represents only one of several potential ways in which to use the LUPM. Therefore, the software must be flexible enough to adapt to different research project requirements as we continue to investigate its application.

How to Use This Report

This report consists of five sections, including the introductory section presented here. The Risk Analysis Using the LUPM section describes a framework for risk analysis involving the LUPM to supplement the analysis. This framework includes defining the study region and hazards, collecting and preparing data, developing mitigation strategies, specifying and evaluating mitigation scenarios, and completing the risk assessment. These steps are covered in detail along with a brief discussion about how the LUPM fits into the risk-analysis process.

The LUPM v1.0 Application Framework section introduces the three packages included with the LUPM v1.0 software. These packages include the PM ArcGIS Extension, the PM User Control, and the LUPM Geoprocessing Tools. Next, the section describes the two LUPM implementations on which the packages are based—the PM Tool and LUPM Geoprocessing Functions. The section concludes with a description of the LUPM core libraries, a set of software modules used to develop the LUPM implementations and/or the LUPM packages.

The Using LUPM v1.0 Packages in the Risk Analysis Process section describes the steps involved in creating an LUPM analysis (that is, scenario) using the LUPM packages. The section begins with the PM Tool-based packages, which include the PM User Control and the PM ArcGIS Extension. Details include setting up the PM Database, selecting features to mitigate, creating hazard-event data, creating and executing scenarios, and evaluating results. Next, the section covers the LUPM Geoprocessing Tools package, with details on creating hazard-event data, selecting features to mitigate, creating and executing scenarios, and using additional geoprocessing tool capabilities.

The LUPM v1.0 Tutorials section presents tutorials about using the PM Tool and the LUPM v1.0 Geoprocessing Tools. The PM Tool tutorial consists of three parts—basic PM Tool operations, scenario development, and scenario results analysis. Part 1: Basic PM Tool operations, includes exercises that walk a user through the steps of creating a PM Database and using the Hazard Events Manager and Scenarios Manager. Part 2: Scenario Development, takes a user through the steps of creating and editing LUPM scenarios using the Add/Edit Scenario dialog. Finally, Part 3: Analyzing Scenario Results, looks at some of the tools for viewing and analyzing scenario results. These tools include the Report Viewer, the Chart tool, and the Comparison tool.

The LUPM v1.0 Geoprocessing Tools tutorial also consists of three parts—running a basic LUPM model, building a custom LUPM model, and additional examples. Part 1, Running the Basic LUPM Model, provides instructions on accessing and running a basic LUPM geoprocessing model—first, without mitigation, and second, with mitigation. Part 2: Building a Custom LUPM Model, walks a user through the steps of creating an LUPM model from scratch using the provided LUPM geoprocessing tools and scripts. Finally, Part 3: Running Additional Examples, briefly discusses additional example models included in the LUPM Geoprocessing Tools package.

References and appendices are included at the end of the report. Appendix A provides a set of instructions for installing the LUPM v1.0 software. Appendix B describes the attributes found in the sample dataset included with the software and used for the tutorial. Appendix C provides documentation for the LUPM v1.0 Geoprocessing Tools. Finally, appendix D describes the LUPM core libraries.

Risk Analysis Using LUPM v1.0

The LUPM v1.0 software implements a scenario-based approach to natural-hazard risk analysis, whereby the outcome of a scenario is influenced by several factors, including the probability of the event occurring, the mitigation strategy (mitigation action and selection of assets to mitigate), and the selection of input data to use for the LUPM v1.0 software. A scenario run produces a report summarizing the effect of the risk-reduction mitigation strategy on the community or region. The report includes avoided losses, retained wealth, return-on-investment, and a measure of the uncertainty associated with the return. Results from multiple scenario runs can be used as a basis for follow-up risk-analysis work (cost-benefit and/or return-on-investment analysis), the findings from which may provide a clearer understanding of the risk-return trade-offs among different risk-reduction strategies.

Risk analysis incorporating the LUPM v1.0 software involves a number of steps, some preceding the use of the software, and some following its use. The basic steps include: defining the study region and hazards, collecting and preparing the data, defining mitigation strategies in the form of mitigation scenarios, executing the LUPM v1.0 software using the scenarios and data specified, and applying the results to complete the risk assessment. LUPM v1.0 fits into this process by enabling the analyst to combine data about assets and hazards into specific mitigation scenarios, based on defined mitigation

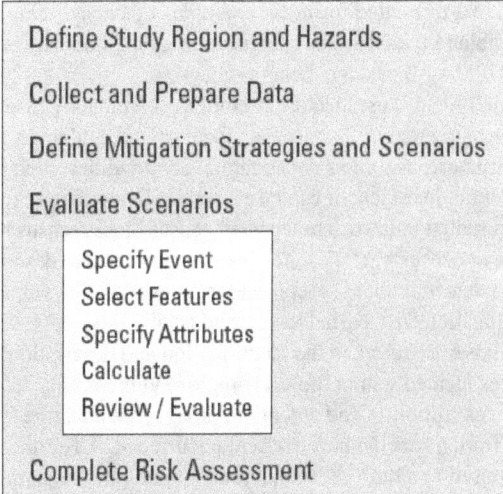

Figure 1. Risk-analysis framework.

strategies to estimate loss avoidance, which may then be used to assess different policy options (fig. 1).

Define Study Region and Hazards

Defining the study region serves to identify the jurisdictional boundaries of the geographical area of interest and to limit the scope of the study to investigate the hazards and related risk-reduction strategies that can affect the communities in the region. A region can be bounded by a county, a State, or some other appropriate boundary. Additionally, all hazards affecting the region, as well as the potential events associated with each hazard, should be identified. This provides the foundation for collecting the data to be used for evaluating mitigation strategies.

Collect and Prepare Data

The next step involves collecting data relevant to the study region and the hazard events that could impact the region. LUPM v1.0 requires two kinds of data—hazard-event data and feature data. The hazard-event data reflects the kinds and probabilities of different hazards of interest. The feature-attribute data involves several different categories of information—exposure, vulnerability, mitigation cost, partial-loss estimates, and ancillary benefits—incorporated into one or more feature layers. The feature data in particular may require a number of preparatory steps and will generally require proficiency with GIS and familiarity using GIS applications, such as ArcMap.

Hazard-event data refers to a potential occurrence of a single natural hazard based on severity and likelihood of the event, and it must be prepared in advance of using LUPM v1.0. Regardless of how the data is prepared, a hazard-event record consists of the hazard-event identifier, name, event probability, and description fields. Hazard-event data will be stored in a formatted form (for example, a database table) accessible to the GIS. The specific storage method depends on the LUPM v1.0 package being used.

LUPM v1.0 uses feature data for the community assets in a given study region. These assets include land, buildings, and infrastructure, as well as nonstructural commodities, such as agriculture, livestock, or capital resources. Every feature represents a spatial unit, such as a parcel, census block, census block group, or census tract. Feature-attribute data, used to describe these assets, fall under several categories—exposure, vulnerability, mitigation cost, partial losses, and ancillary benefits.

Exposure refers to the identification and description of all features located within the study region. Vulnerability refers to the susceptibility and extent of damage that a feature may incur from a specific hazard-event occurrence. A feature may be exposed to a hazard event, but its vulnerability to damage depends largely on its level of readiness to withstand the impacts from such an event. For example, an unreinforced building is more likely than a structurally-upgraded building to sustain damage from a moderately strong earthquake.

Mitigation cost refers to the dollar amount invested to reduce an asset's risk of loss from a specific hazard event. Mitigation applied at the feature level may be referred to as site mitigation, whereas mitigation applied at a regional level may be referred to as community mitigation. LUPM v1.0 handles site mitigation using the mitigation-cost data assigned to each feature and handles community mitigation using an aggregated value representing the costs to mitigate assets serving the entire region (for example, essential facilities such as hospitals, facilities facing high potential loss, and transportation and utility systems).

Partial losses and ancillary benefit data are optional inputs for LUPM v1.0. Partial losses are estimates measuring the extent of loss that a feature could incur, if mitigated and not mitigated, respectively. By default, LUPM v1.0 assumes a feature will incur no loss if mitigated, and complete loss if not mitigated. Partial-loss estimates enable LUPM v1.0 to assume that mitigation can be less than 100 percent effective against protecting a feature from damage and that a feature can incur less than 100 percent loss even without mitigation. Ancillary benefits refer to nonmonetary benefits or characteristics that a feature may have or provide. LUPM v1.0 recognizes five types of ancillary benefits—cultural, critical, economic, environmental, and safety. A benefit value of 1 indicates that a feature provides that benefit, whereas a value of 0 indicates that it does not.

Collecting data for hazard-mitigation analysis is a difficult, expensive, and time-consuming task, often requiring assistance from individuals or organizations having the scientific, mathematical, or engineering expertise and resources required to produce such data. The discipline involving the research and development of methods for estimating damages and losses from natural-hazard events has been growing due to demand for such data. FEMA's HAZUS-MH is a possible source for data to use with LUPM v1.0.

HAZUS-MH provides a regional, scenario-based approach to estimating economic losses incurred from an occurrence of an earthquake-, flood-, or hurricane-hazard event (Buriks, 2004).

HAZUS-MH includes a base set of inventory data, such as exposure data, including buildings, infrastructures, and demographics, covering most of the United States. HAZUS-MH may be used to generate a scenario from the base data provided or from user-supplied data, different loss-estimate equations, or both. The output from a HAZUS-MH scenario run can be subsequently used to derive data for LUPM v1.0. Although this approach has some limitations, it provides a way of producing data for LUPM v1.0 when such data are not readily available. The sample dataset for the tutorial provided in this report was derived using the output from a HAZUS-MH generated scenario (see appendix B).

Develop Mitigation Strategies

At this stage, the analyst is prepared to plan and formulate mitigation strategies for each hazard affecting the study region in order to provide a basis for comparing and evaluating mitigation alternatives. This process involves making important decisions about assets (and possibly locations or areas) to protect and what mitigation actions to take. These decisions are influenced by a number of factors, including perceptions about the risk of hazards and the tolerances towards risk, budget and policy constraints, planning and development objectives, social and economic impacts, and mitigation priorities (Wein, 2007). A perfect risk-reduction strategy would be difficult, even impossible, to derive. Rather, the analyst can weigh risk-return benefits and tradeoffs by developing multiple risk-reduction strategies to make a more informed assessment of alternative risk-reduction strategies.

As an example, let's assume a hazard event involving a magnitude M7.0 earthquake along the Hayward Fault with a 30 percent chance, on average, of occurring within the next 100 years. This fault is located just east of San Francisco and extends along most of the San Francisco Bay Area (fig. 2). Let's also assume that our mitigation strategy places a priority on protecting people living in homes classified as light wood-framed, single-family residential structures. Given these assumptions, one mitigation strategy would be to upgrade structures of this type to meet a more stringent building code that would also offer more protection against an earthquake. Budget constraints, however, would make applying this mitigation action to every structure unrealistic. Therefore, a strategy to limit the number of structures to mitigate can be based on other criteria, such as proximity to the fault.

Ideally, alternative mitigation strategies for each hazard event should be developed. For example, one mitigation option would provide relocation assistance to people currently living in structures close to a hazard, such as the Hayward Fault. This action may prove to be more cost effective than a plan to upgrade structures. Another mitigation strategy would be to place a greater emphasis on response and recovery, rather than planning and prevention efforts. This option can be implemented by using an assortment of measures, including insurance, temporary housing, and other programs designed to assist people who are displaced from their homes due to an occurrence of the event. The analyst can better understand the risk-return

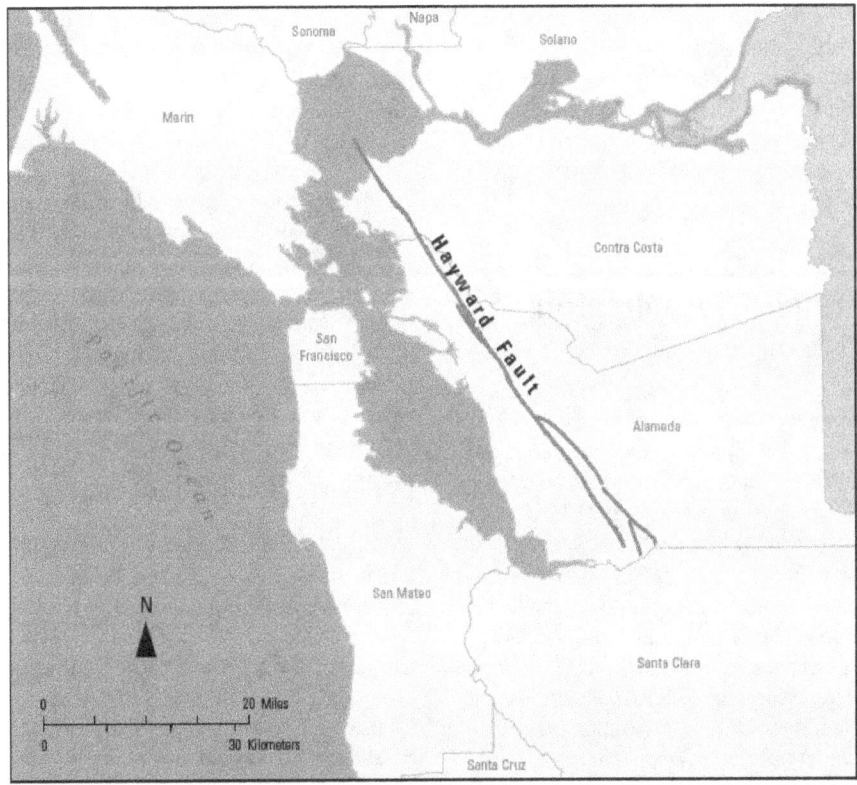

Figure 2. San Francisco Bay Area, California, showing the Hayward Fault.

benefits and trade-offs of one strategy over another by comparing the results of these and other mitigation strategies.

Specify and Evaluate Mitigation Scenarios

LUPM v1.0 is aimed at evaluating mitigation strategies—in particular, LUPM v1.0 is designed to evaluate scenarios derived from mitigation strategies. For example, a mitigation strategy may place a priority on protecting the locations that represent the highest value. A number of scenarios characterizing this strategy may be developed, representing different numbers of locations actually mitigated (for example, the top 5 percent of locations by value rather than the top 10 percent).

Thus, an LUPM v1.0 scenario is the specific implementation of a mitigation strategy. It includes the selection of a specific hazard event (for example, a M7.0 magnitude earthquake), the geographic features being analyzed (for example, parcels), the attributes for these features (value, susceptibility to damage, cost of the mitigation, etc.), and the selection of features to be protected through a mitigation effort.

Returning to the Hayward Fault example described in the previous section, we can develop a number of mitigation scenarios based on a strategy to protect assets close to the fault. We can vary the distance (for example, all structures located within 3, 5, and 8 kilometers of the fault). We can vary this scenario even further by specifying locations that meet different value thresholds, such as all structures whose value

is below $300,000. These scenarios may then be compared to assess how they impact community wealth.

The analyst uses the LUPM v1.0 tools to model each of the mitigation scenarios. Each scenario requires specifying 1) a hazard event, 2) selection of features to mitigate, and 3) inputs to use (that is, data describing asset value, mitigation cost, damage susceptibility). The results of a scenario run provide an indication of the losses that may be avoided through mitigation and, therefore, the impact on community wealth. The results can also be saved and compared with results from other scenarios.

Complete Risk Assessment

The final step of the risk analysis involves using the results from each of the scenario runs as a source for follow-up work to complete the risk analysis and assessment. An example of such work includes a cost-benefit analysis comparing the costs and benefits (avoided losses or impact on community wealth) of different strategies for different hazard events. Another example is a risk-return analysis of the trade-off between the risks (financial and effective risks) of investing in mitigation and the return (losses avoided relative to mitigation cost) mitigation can provide. Such analyses also provide a basis for comparing and ranking risk-reduction policies for different hazard events. These comparisons can then help to prioritize the mitigation of multiple events based, perhaps, on the frequency of each event in relation to the risk-return tradeoffs and benefits achieved through various mitigation strategies.

The LUPM v1.0 software supports and supplements risk analysis; it does not replace it. The tool should be used within a context of a broader risk-analysis framework. Employing such a framework will make best use of the benefits offered by the LUPM v1.0 tools.

The LUPM Version 1.0 Application Framework

The LUPM v1.0 software architecture is organized into a series of major components to support analysis of mitigation scenarios (fig. 3). LUPM v1.0 functions are accessed by using one of three software packages—a user control (PM User Control), an ArcGIS extension (PM ArcGIS Extension), or a set of ArcGIS geoprocessing tools (LUPM v1.0 Geoprocessing Tools). These packages connect the event data, the feature data, and the features selected for mitigation with the core LUPM calculations. Each package provides methods to identify the relevant data, to perform the LUPM calculations, and to display and save the results. Ultimately, each package can be used to support a variety of research applications.

The LUPM v1.0 packages are based on one of two implementations—the PM Tool and the LUPM v1.0 Geoprocessing Functions. The implementations provide access to the underlying LUPM v1.0 functionality for different processing environments. The PM User Control and the PM ArcGIS Extension are based on the PM Tool, and the LUPM v1.0 Geoprocessing Tools are based on the LUPM v1.0 Geoprocessing Functions.

Finally, the packages and implementations are built on top of the LUPM v1.0 core libraries, including the PM_Math, PM_Common, and PM_ESRI libraries. The libraries perform

the calculations, provide tools to help prepare the data for analysis, and provide access to ArcGIS functions to support the LUPM. This ensures that all of the packages employ the same calculations.

The LUPM v1.0 software processes feature data through an ArcGIS feature layer. A feature layer provides a visual representation of a collection of geographic features sharing the same geometry type (point, line, or polygon), attributes, and spatial reference. A layer does not actually contain the data, but, rather, it makes a reference to feature class (an actual ArcGIS data source, such as a shapefile, coverage, or geodatabase feature table). LUPM v1.0's use of the feature layer is important because selections are stored for feature layers, not feature classes, and selection is the means by which the user identifies which features are to be mitigated.

LUPM Version 1.0 Packages

The LUPM v1.0 packages provide different ways for an analyst to evaluate a mitigation scenario. The PM ArcGIS Extension enables the analyst to perform the evaluation within a self-contained environment in the ArcMap application environment. The hazard-event data creation, mitigation-scenario definition and execution, and LUPM result examination are all handled within the tool. The PM User Control provides similar capabilities in a standalone application environment, or embedded in another application. Finally, the LUPM v1.0 Geoprocessing Tools enable the analyst to perform LUPM calculations in the ArcGIS Geoprocessing environment using standard and custom ArcGIS tools, as well as the LUPM v1.0 tools.

The PM ArcGIS Extension Package

The PM ArcGIS Extension is an ArcGIS extension control. An ArcGIS extension is a custom software component providing added functionality to a particular ArcGIS desktop application. In this case, the PM ArcGIS Extension consists of a single tool whose corresponding icon is added to the ArcMap desktop application's toolbar (fig. 4). The extension's primary function is to provide a link between the ArcMap desktop application and the PM Tool, permitting the tool to use the active ArcMap document as input.

The PM User Control Package

The PM User Control is a Microsoft .NET Windows user control—a reusable software component that includes a graphical user interface and performs a particular function. Specifically, the PM User Control encapsulates a map interface, enabling a user to view, edit, and perform other GIS operations on ArcGIS-based data sources, such as an ArcMap document. The interface includes a toolbar with an icon that links to the PM Tool. The link between the tool and the map interface allows the tool to access and process the contents of the active ArcMap document. The control's mapping capabilities were programmed using components from the ArcGIS Engine software developer kit—a toolkit used by application

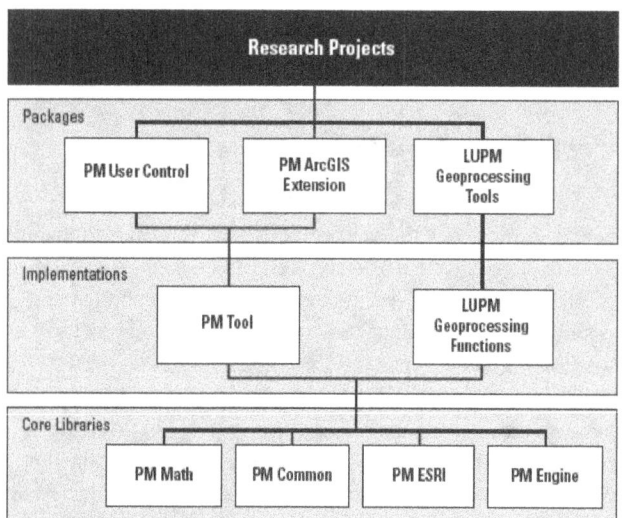

Figure 3. Chart showing Land Use Portfolio Modeler version 1.0 software architecture.

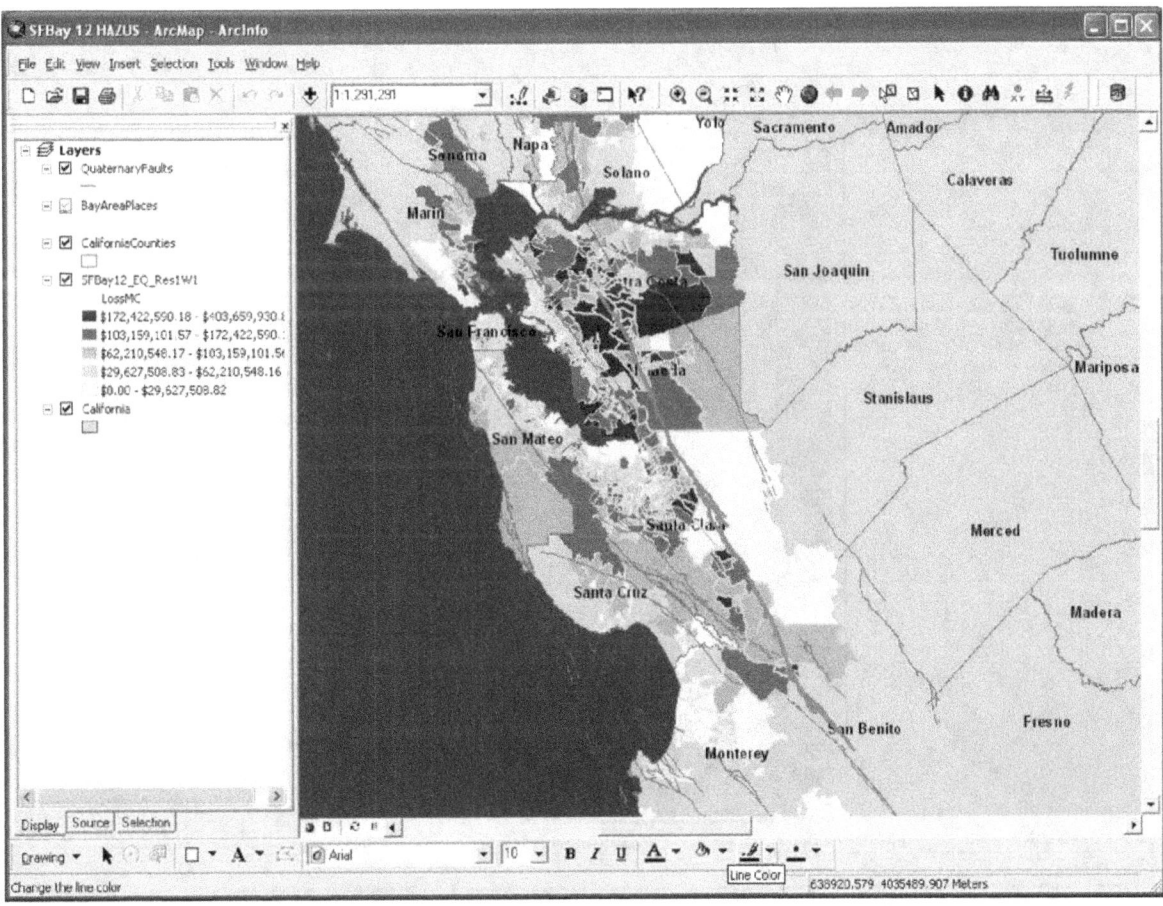

Figure 4. Screen shot showing the PM ArcGIS Extension docked in ArcMap toolbar.

developers to build and deploy custom GIS and mapping applications. The PM User Control is embedded in either a desktop Windows application or another Windows component module (fig. 5). A sample application using the control is included with the LUPM v1.0 software.

The LUPM v1.0 Geoprocessing Tools Package

The LUPM v1.0 Geoprocessing Tools package provides an implementation of LUPM v1.0 that runs within the ArcGIS ModelBuilder environment. The tools are a set of geoprocessing tools and associated Python scripts that prepare the data for the LUPM calculation, execute the calculation, and handle the results. The tools may be used in various combinations in a geoprocessing model for the ArcGIS ModelBuilder. The package is contained in USGS LUPM Toolbox, an ArcGIS toolbox.

LUPM Implementations

The LUPM v1.0 implementations provide the foundations for invoking the LUPM in different application environments. The PM Tool-based packages are self-contained applications that provide all of the tools necessary to use the LUPM to

analyze mitigation scenarios. The LUPM v1.0 Geoprocessing Functions enable the analysis of mitigation scenarios within the ArcGIS ModelBuilder environment. Both of these implementations use the same underlying calculations provided by the LUPM v1.0 core libraries.

The PM Tool Implementation

The PM Tool provides a forms-driven modeling environment, utilizing a set of graphical user interfaces from one of the core libraries called PM_UI. The tool implements the LUPM using a scenario-based approach to modeling a given mitigation strategy. Each run of the model is based on a scenario specifying a particular hazard event and mitigation choices based on a mitigation strategy. A mitigation strategy is a selection of features to mitigate and data characterizing those features. The PM Tool uses the active ArcMap document to identify the feature layers, determine which features from those layers were selected for mitigation, and the attribute fields containing the data to use for the model run.

The PM Tool is accessed through a link provided in either the PM User Control or the PM ArcGIS Extension. Launching the tool will display the tool's primary dialog window, the PM

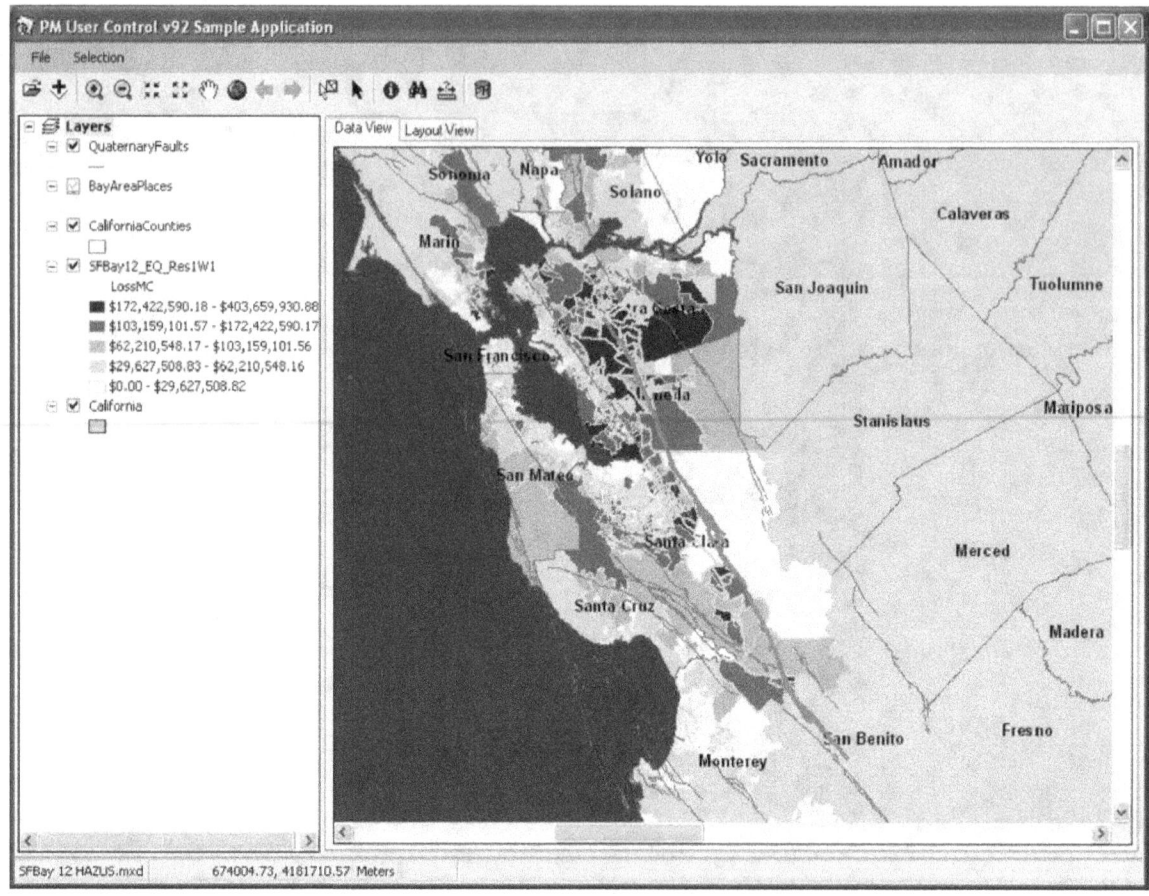

Figure 5. Screen shot showing the PM User Control dialog window embedded in a desktop application.

Tool dialog window (fig. 6). This window provides access to the tool's functions, including Database Setup, Hazard Events Manager, and Scenarios Manager, as well as References and Help.

The PM Database is a customized Microsoft Access database used by the PM Tool and accessed from either the PM ArcGIS Extension or the PM User Control package. This database is used for storing hazard and scenario data created by using the tool. All hazard and scenario data are managed by using the PM Tool software, but the tables are accessible as standard Microsoft Access tables. Appendix E provides a description of the table objects contained in a PM Database.

After launching the PM Tool, a user must connect to a PM Database prior to performing activities, such as creating hazard events or scenarios. The tool's Database Setup dialog is used for creating a new PM Database or connecting to an existing database. Any number of PM Databases can be created, but only one database can be used at any given time while the tool is running.

The LUPM v1.0 Geoprocessing Functions Implementation

The LUPM v1.0 Geoprocessing Functions implementation contains a set of custom geoprocessing function tools that access

the LUPM calculations in the PM_Math core library (usgs-FunctionLUPMCalculate) and ArcGIS capabilities that are not accessible to scripts written at the Python level, such as returning a feature layer in XML format (usgsFunctionGetFeatureLayer-AsXML). These tools are incorporated into the Python script tools that make up the LUPM v1.0 Geoprocessing Tools package. They, along with standard ArcGIS geoprocessing tools, enable the analyst to assemble a set of data, operations, and parameters into a model for the ArcGIS ModelBuilder environment.

LUPM Core Libraries

The foundation of the LUPM v1.0 software architecture is comprised of a set of core library components or libraries, which provide the necessary building blocks for developing tools and applications implementing the LUPM. Each of the three packages included with the LUPM v1.0 software was implemented by using a number of these core libraries. At a minimum, all three implementations require the PM_Math, PM_Common, and PM_ESRI libraries.

The PM_Math library contains the logic used to implement the LUPM, providing functions that prepare the LUPM input data, perform the LUPM calculations, and return results in various

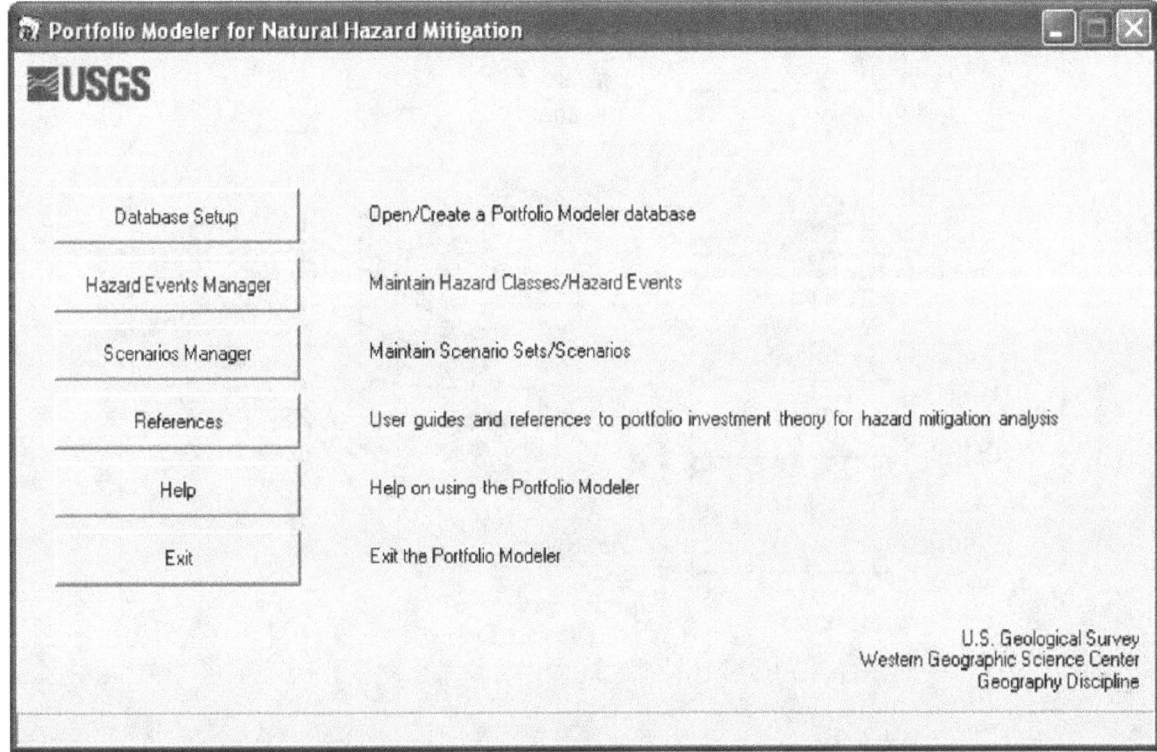

Figure 6. Screen shot showing the PM Tool dialog window.

formats. The PM_Common library contains the definitions of various data structure class objects, which serve as containers for data and are made use of throughout the core set of libraries. Finally, the PM_ESRI library provides a number of utility functions involving the manipulation or processing of ESRI-based objects, such as retrieving data from an ArcMap feature layer. Appendix D provides a table summarizing all the LUPM v1.0 core libraries.

The core library components were written using the C# programming language. These components are based on the Microsoft.NET framework, version 2.0. Additionally, the PM_ Engine core library uses components from the ArcGIS Engine software developer kit. This toolkit provides the components needed to build custom mapping applications or tools with ArcGIS functionality.

Using the LUPM v1.0 Packages to Evaluate Mitigation Scenarios

This section provides details on applying the LUPM v1.0 packages for risk analysis, specifically focusing on the flow of activity typically followed when using a package. These activities include selecting features to mitigate, defining hazard-event data, building and executing scenarios, and evaluating results.

Creating Scenarios with PM Tool-based Packages

A user prepares to use either PM Tool-based package by first launching the appropriate mapping application. The PM ArcGIS Extension is run in ArcMap, whereas the PM User Control package is run from the Windows desktop program in which it is embedded. The PM User Control includes its own map interface providing basic mapping functionality to interact with ArcMap documents and other ArcGIS-based data sources. The process then becomes the same for either PM Tool-based package once the applications are run and the PM Tool is accessed (fig. 7).

The PM Tool requires two sources of data—an ArcMap document and a PM Database. The user opens an existing ArcMap document containing necessary feature layers or loads the individual feature data layers into an active ArcMap document. A reference to the active ArcMap document is passed to the tool when it is launched from either the PM User Control or the ArcMap application using the PM ArcGIS Extension. Access to a PM Database, however, is set only after the PM Tool has been launched.

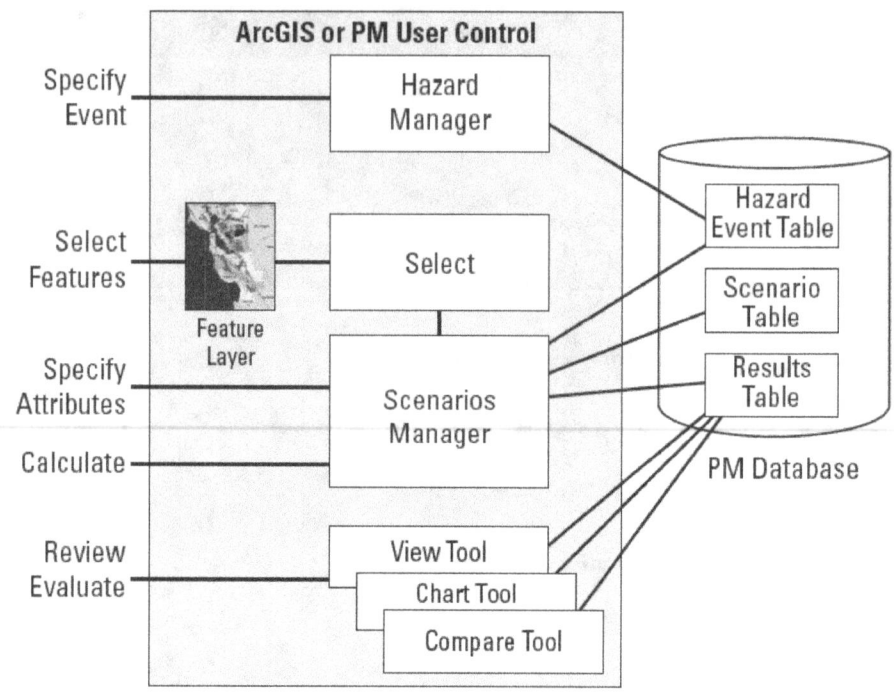

Figure 7.

Set Up the PM Database

The Portfolio Modeler-Database Setup dialog is used for either selecting an existing PM Database or creating a new one (fig. 8). An existing PM Database can be found using the Browse function, which returns the location and name of the database to the corresponding database text box. A new PM Database is created using the Create function, which creates a database and returns its location and name to the database text box. Clicking the OK button will select the database and exit the Portfolio Modeler —Database Setup dialog. Selecting the database merely returns the name and location of the database to the tool, identifying it as the database against which transactions will be performed. The database is only connected to and accessed as needed such as when a transaction is to be committed (that is, saved) to the database.

Select Features to Mitigate

Selecting features for mitigation is performed outside of the PM Tool environment using standard ArcGIS feature selection tools. The selection is based on the defined criteria for a given mitigation strategy. No selection is required if the strategy calls for no site mitigation. Selection may use any combination of the Select By Attributes and Select By Location queries. Select By Attributes is used to select features from a given layer based on criteria defined in a Structured Query Language (SQL) query statement. Select By Location is used to select features from a given layer based on their spatial relation to other features within the same layer or other layers. Details on how these tools are used are covered in the tutorial, as well as in the ArcGIS documentation.

Figure 8.   Screen shot of the Portfolio Modeler - Database Setup dialog window.

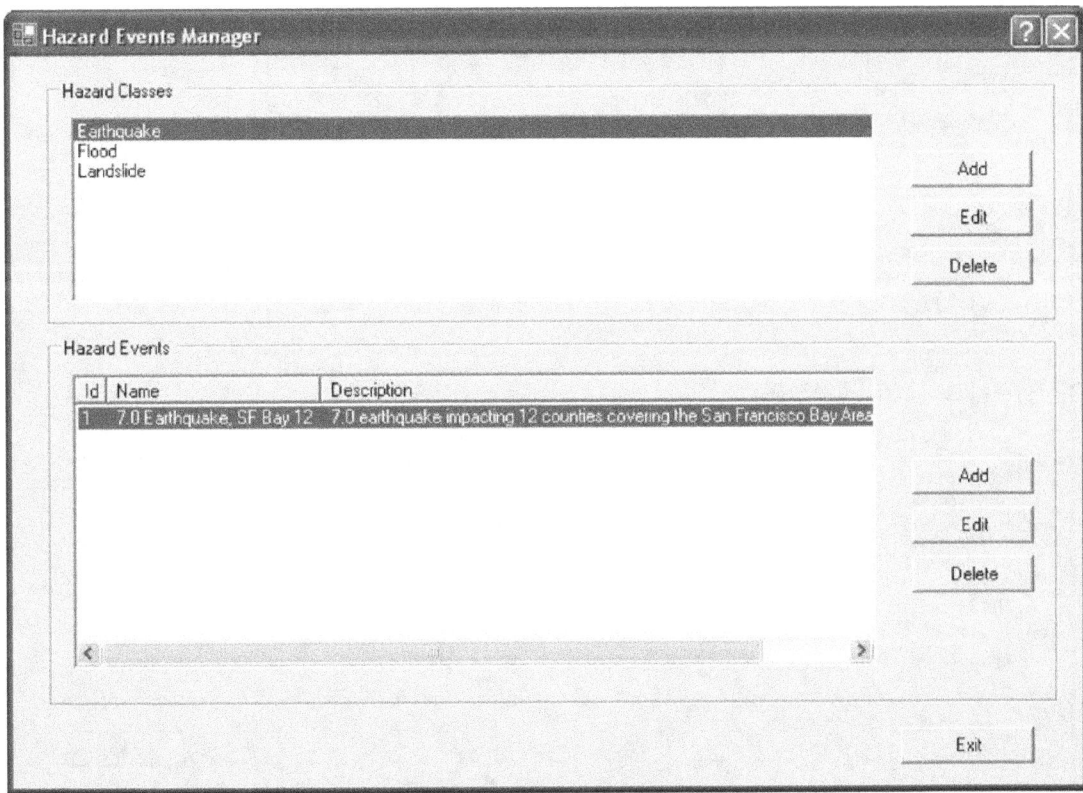

Figure 9. Screen shot showing the Hazard Events Manager dialog window.

Create Hazard-Events Data—The Hazard Events Manager

The Hazard Events Manager (fig. 9) provides a dialog for defining and managing hazard classes and hazard events. A hazard class describes a type of hazard, such as an earthquake, whereas a hazard event represents an instance of a particular hazard class (for example, a 7.0 Earthquake). Hazard classes and events may be added, edited, or deleted using their respective functions from the dialog. A hazard event is associated with a specific hazard class, so the class must exist before the event is created. For instance, the user must first create an Earthquake hazard class, if it does not already exist, before creating the-earthquake hazard event.

Hazard classes are added or edited using the Add/Edit Hazard Class dialog window, which provides entry fields for the name and description of a hazard class. Similarly, hazard events are added or edited using the Add/Edit Hazard Event dialog window, which provides entry fields for the name, description, and probability of a hazard event. Hazard classes and events are deleted using their respective Delete functions, which will prompt the user to confirm the action before committing the delete transaction. Database integrity rules are enforced during a delete transaction, whereby deleting a hazard class will also delete all hazard events associated with it. Additionally, a hazard event can only be deleted if no scenarios are associated with that event. In that case, all the scenarios associated with the hazard event must be deleted before the event can be deleted.

Create and Execute Scenarios—The Scenarios Manager

The Scenarios Manager provides a dialog for creating and managing scenario sets and scenarios (fig. 10). A scenario set organizes scenarios by collecting scenarios for a particular hazard event in one place. A scenario represents an instance of an LUPM v1.0 run, which is based on a specified mitigation strategy for a specific hazard event. Scenario sets and scenarios may be added, edited, or deleted using the corresponding Add, Edit, and Delete buttons found in their respective sections of the dialog window. The dialog window also includes buttons to access several tools for reporting scenario results, including View, Chart, and Compare. The View button displays the model output for a selected scenario. The Chart button links to a tool for producing charts using selected model results for one or more scenarios. The Compare button launches a tool that produces a cross-matrix report, providing a comparison of outputs for one or more scenarios.

A scenario is associated with a specific scenario set, much like a hazard event must be associated with a specific hazard class. Hence, a scenario set must exist before creating

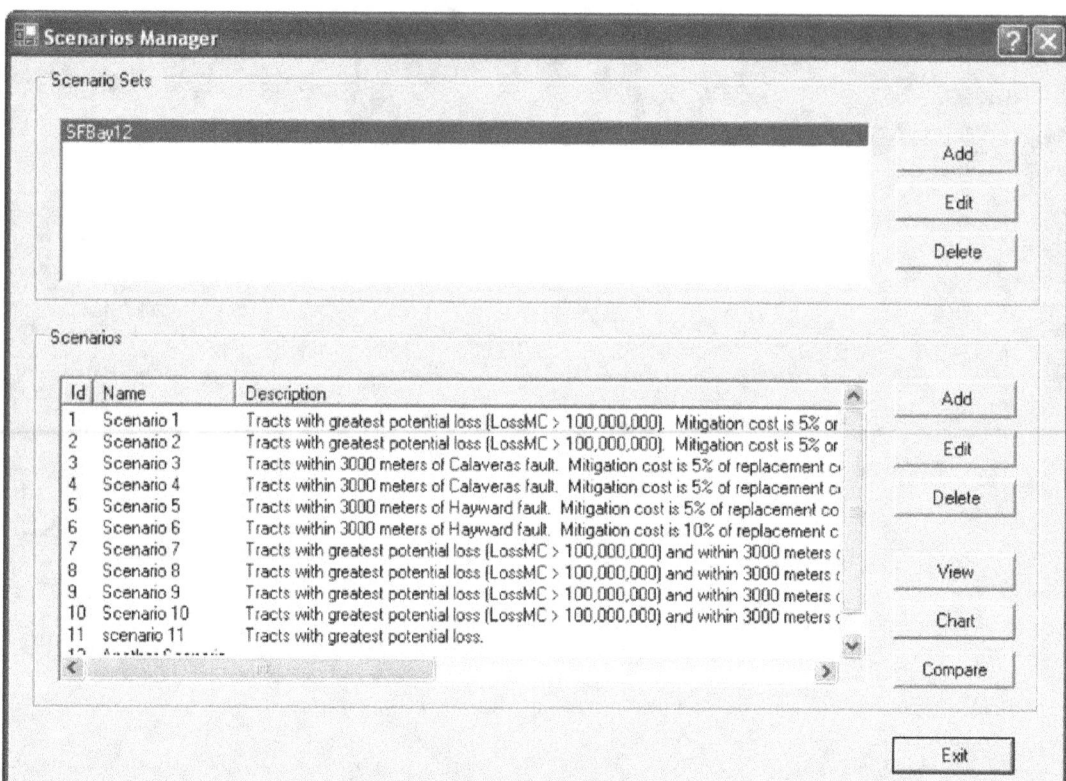

Figure 10.   Screen shot showing the Scenarios Manager dialog window.

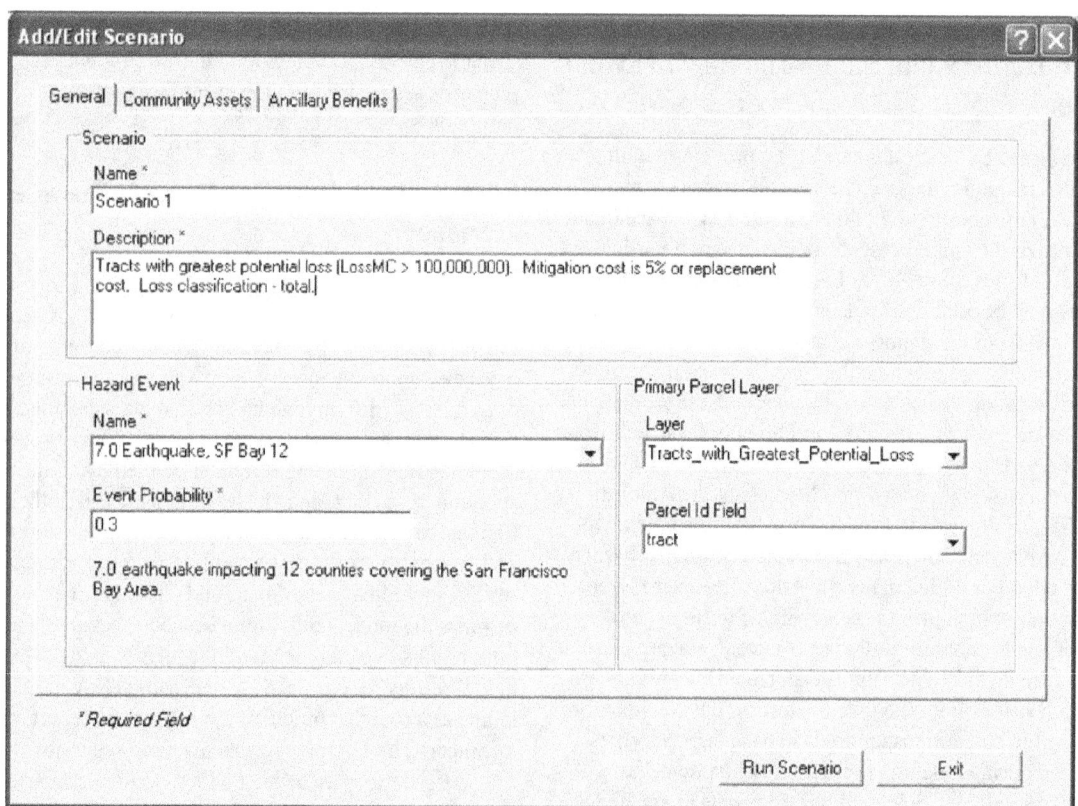

Figure 11.   Screen shot of the Add/Edit Scenario dialog—General tab page.

scenarios that are to be associated with it. A scenario set may include many different scenarios. As a practical matter, however, the user should use scenario sets systematically. For example, a user may wish to create a scenario set for a specific hazard event, and then associate all scenarios based on that hazard event with the scenario set.

Scenario sets are added or edited using the Add/Edit Scenario Set dialog window, which provides entry fields for the name and description of a scenario set. Likewise, scenarios are added or edited using the Add/Edit Scenario dialog window. To add a scenario, the user selects a scenario set, which will contain the new scenario, and clicks the Add button located in the Scenarios section. To edit a scenario, the user selects the scenario set containing the scenario and then the scenario itself and clicks the Edit button located in the Scenarios section. Scenarios sets and scenarios are deleted using their respective Delete functions, which will prompt the user to confirm the action before committing the delete transaction. Database integrity rules are enforced during a delete transaction, whereby deleting a scenario set will also delete all scenarios associated with it.

Adding a scenario: The Add/Edit Scenario Dialog Window

The Add/Edit Scenario dialog window contains entry fields located in three tabs—General, Community Assets, and Ancillary Benefits. The General tab page (fig. 11) provides fields used to specify the scenario (name and description), the hazard event, and, optionally, the primary data layer. The hazard event is selected from the Name drop-down list. When selected, the probability for the chosen event is automatically entered in the Event Probability field. This value can be modified, if desired. The Primary Parcel Layer field identifies a layer containing most, if not all, of the fields, which can be used as input to LUPM v1.0. When this layer is specified, the Primary Id field for that layer must also be specified. The Primary Id is a unique value identifying each feature in the layer. Information entered here automatically provides many initial values for similar fields in the Community Assets and Ancillary Benefits tabs.

The Community Assets tab page (fig. 12) contains entry fields used for specifying the layers and fields to use for retrieving feature data describing LUPM inputs, such as asset value, damage susceptibility, mitigation cost, and, optionally, estimates of partial losses. Each of these inputs is specified using three parameters—the feature layer containing the input field to use, the Primary Id field of that layer, and the input field itself. This tab page also contains another optional field—Other Mitigation Costs—used to specify costs for the mitigation of regional assets (for example, highways, dams, bridges, or other assets that affect or have an impact across the study region).

Data in the Assets and Mitigation Cost fields are expressed in dollars, while data for the Susceptibility and the partial-loss estimate fields (Extent Loss Without Mitigation and Extent Loss With Mitigation) are expressed as a fractional value between 0 and 1, inclusive. Partial-loss estimate fields indicate the extent of loss that a feature could incur, both with and without mitigation. If these parameters are left blank, the

LUPM v1.0 software will assume that a feature will incur complete loss if not mitigated and no loss if mitigated.

A feature may contain several fields from which to choose to describe a certain LUPM input, such as mitigation cost. The choice of which fields to use is driven by assumptions that are made and applied to the scenario being modeled. For example, a feature may have a field that describes mitigation cost as being five percent of replacement cost and another field that describes the cost as being ten percent of replacement cost, each representing a different cost assumption. The user would select an appropriate replacement-cost field to represent this input, depending on which assumption is appropriate for that scenario.

The Ancillary Benefits tab page (fig. 13) contains entry fields used for specifying the layers and fields to use for retrieving data associated with ancillary benefit inputs. Ancillary benefits consist of five categories—critical, cultural, economic, environmental, and safety. The specification of any of the ancillary benefit inputs is optional. Ancillary benefits are characteristics that a feature may have or provide. A value of 1 indicates a feature provides that benefit, whereas a 0 value indicates that it does not. Again, three pieces of data are required for each ancillary-benefit parameter—the layer containing the corresponding ancillary-benefit field, the layer's Primary Id field, and the ancillary benefit field itself.

Executing the scenario

The user runs the LUPM v1.0 software on a scenario by clicking the Run Scenario button, which opens a progress window showing the scenario's runtime status. A scenario may be stopped at any time during its run by clicking the Cancel button in the progress window. A Portfolio Modeler Scenario Summary Report is generated for the scenario and displayed in the Report Viewer after the run has completed (fig. 14).

The Portfolio Modeler Scenario Summary Report consists of four parts. The first part of the report provides a description of the scenario, the hazard event, and input parameters used for the retrieval of community-asset and ancillary-benefit data. The second part (Community Wealth) includes information on the total number of features, the number of features mitigated, the original wealth (total assets plus mitigation investment), and the mitigated wealth. The third part of the report displays output for two forms of analysis—simple and probabilistic. The simple analysis applies to the case where the hazard event is assumed to occur. Therefore, the calculations do not use the hazard-event probability. The probabilistic analysis applies to the case where the occurrence of the hazard event is uncertain. Hence, calculations used in this analysis do use the hazard-event probability. The variables reported on, in either analysis, include losses with and without mitigation, avoided losses, retained wealth, and return-on-investment. The probabilistic analysis also reports the uncertainty associated with each variable.

Finally, the fourth part of the report provides a summary of ancillary benefits. Only ancillary benefits whose parameters were specified are included in this summary. Each reported ancillary benefit summarizes the total number of features with

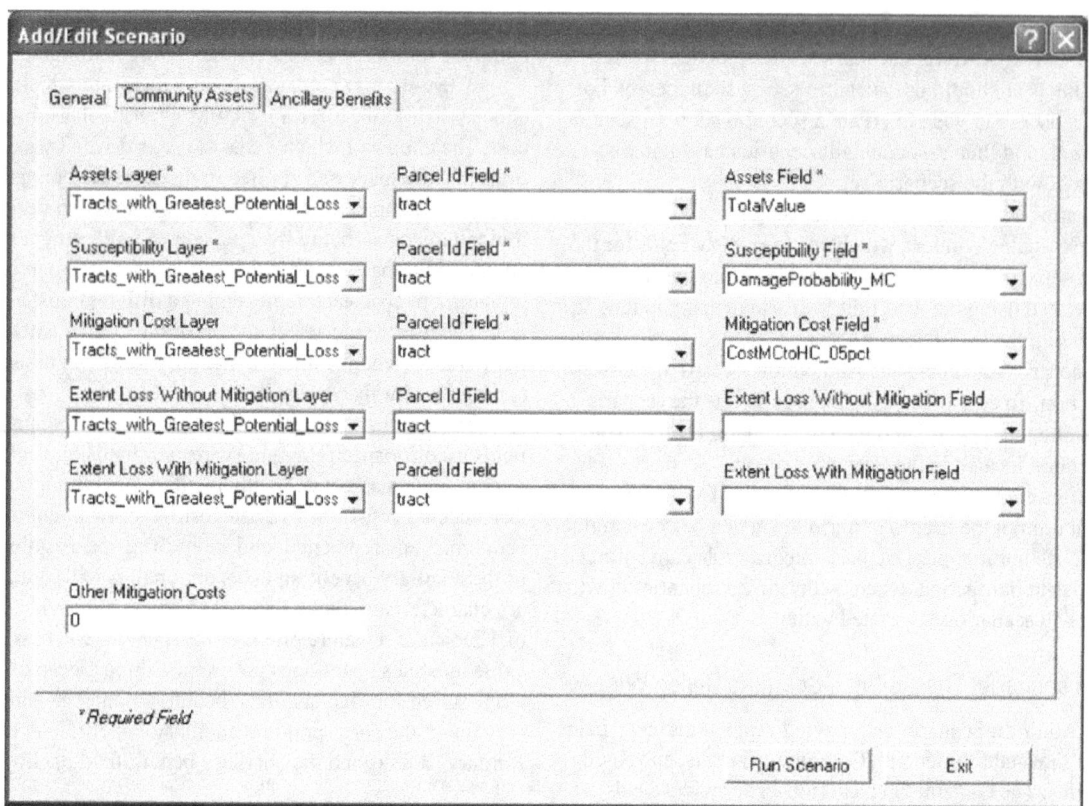

Figure 12.   Screen shot of the Add/Edit Scenario dialog—Community Assets tab page.

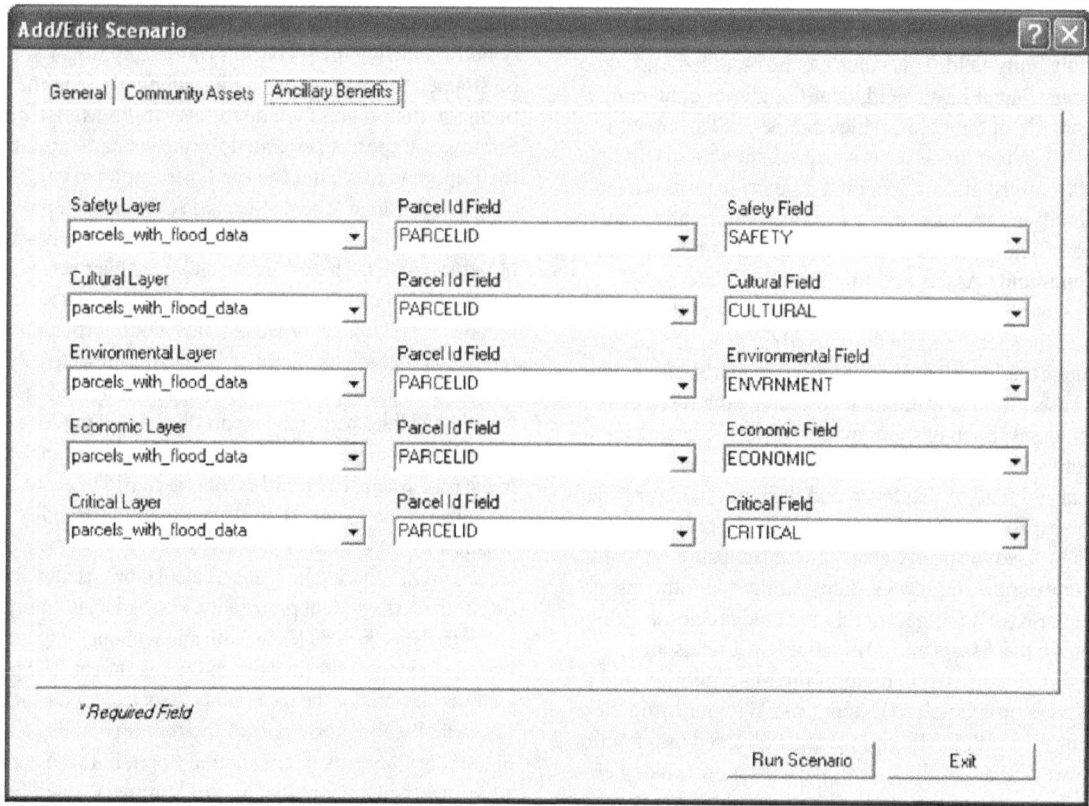

Figure 13.   Screen shot of the Add/Edit Scenario dialog—Ancillary Benefits tab page.

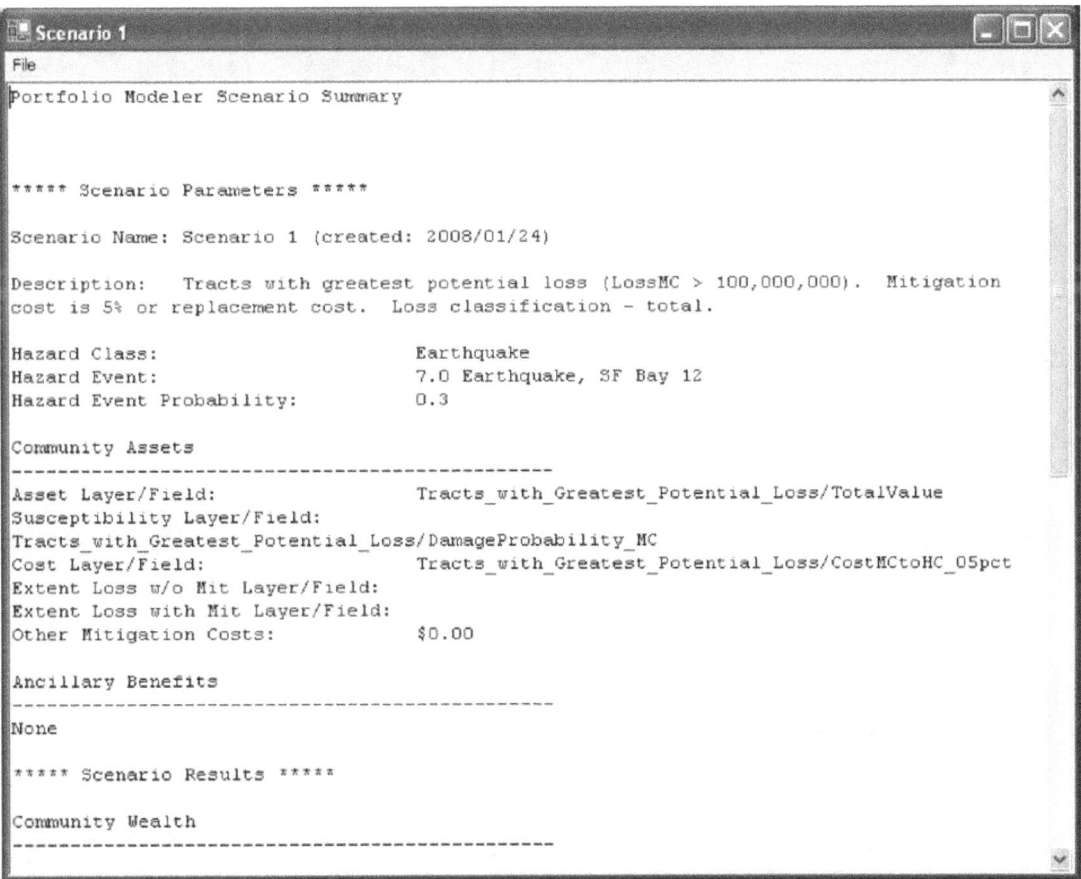

Figure 14. Screen shot of the Portfolio Modeler Scenario Summary Report shown in the Report Viewer dialog.

that benefit versus the number of features with that benefit that were mitigated.

The Report Viewer includes functions for printing and saving the report to a text file. When exiting the viewer for a newly generated scenario, the user will be prompted to choose whether or not to save the scenario to the active PM Database. The user can also choose to save the detailed feature output data for the scenario into a newly created table in the database.

The user may edit and rerun any existing scenario. The edited scenario can replace the existing scenario, or be saved as a new scenario. The user can perform this editing process repetitively to develop a number of scenarios for a given event, applying different assumptions and/or inputs for each scenario as desired. The user can then determine which scenarios provide more favorable results in terms of maximizing avoided losses or yielding a higher return on investment by comparing scenarios to one another.

Evaluate Results—View, Chart, and Compare Tools

The Scenarios Manager dialog includes several scenario output-reporting tools, including View, Chart, and Compare. The View tool retrieves the Portfolio Modeler Scenario Summary Report for a selected scenario, displaying the report in the Report Viewer. The Chart tool can be used to create one of two chart types—a multi-series standard-error chart, or an ancillary-benefit comparison chart. The multi-series standard-error chart displays any two scenario numeric-output fields (such as mitigation cost or retained wealth) and their corresponding standard-deviation fields, if applicable (fig. 15). The ancillary-benefit comparison chart displays any of the available ancillary-benefit fields—critical, cultural, economic, environmental, and safety (fig. 16). Either chart uses output from one or more selected scenarios. Finally, the Compare tool generates a cross-matrix report displaying the portfolio model results for up to four scenarios. Each scenario is reported in a column running along the horizontal axis of the report, with the output displayed in rows running along the vertical axis. The report is displayed in the Scenario Comparison Report Viewer (fig. 17).

Creating Scenarios with the LUPM v1.0 Geoprocessing Tools

The LUPM v1.0 Geoprocessing Tools package provides an ArcGIS ModelBuilder-based implementation of the LUPM. Models created with these tools enable the analyst to associate

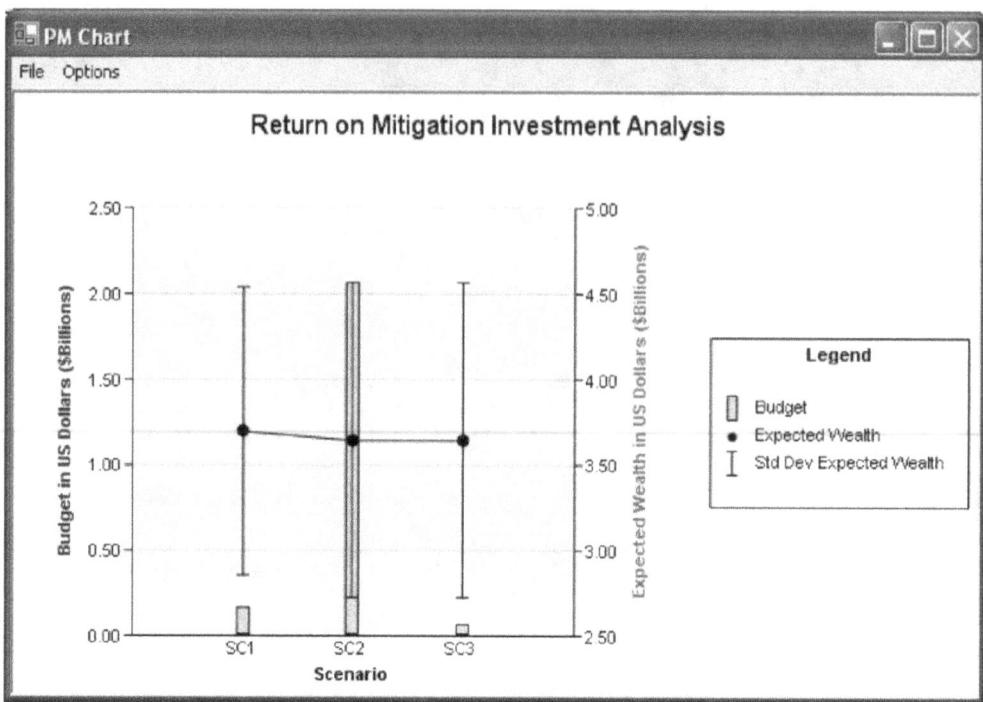

Figure 15. Screen shot of a multiseries standard-error chart.

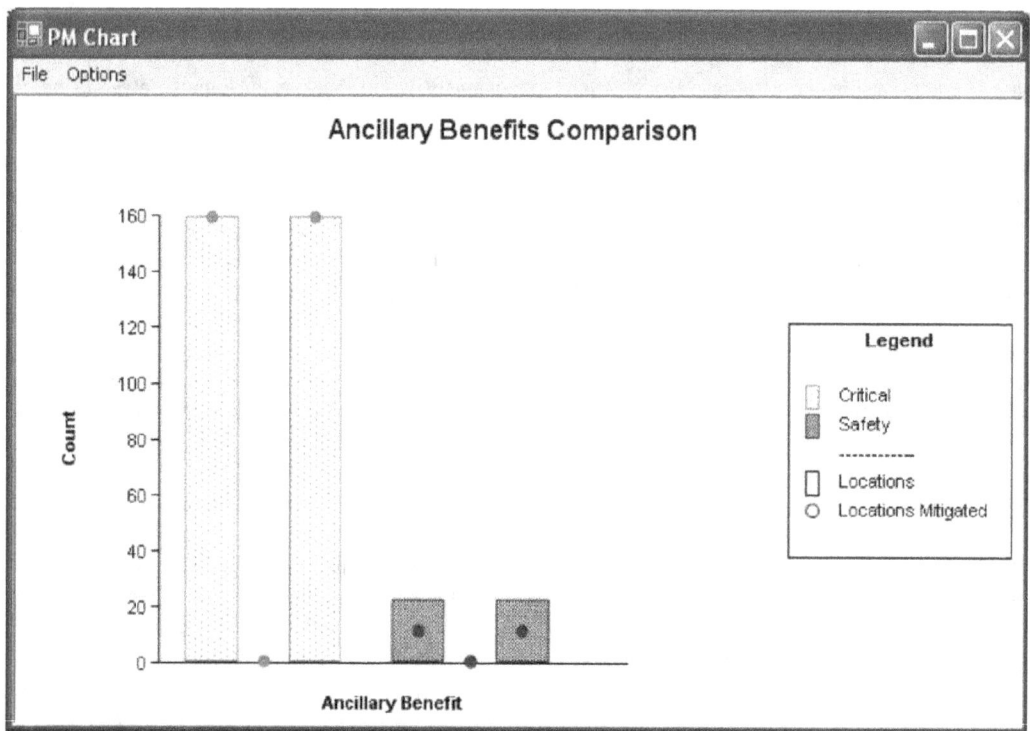

Figure 16. Screen shot of an ancillary-benefit comparison chart.

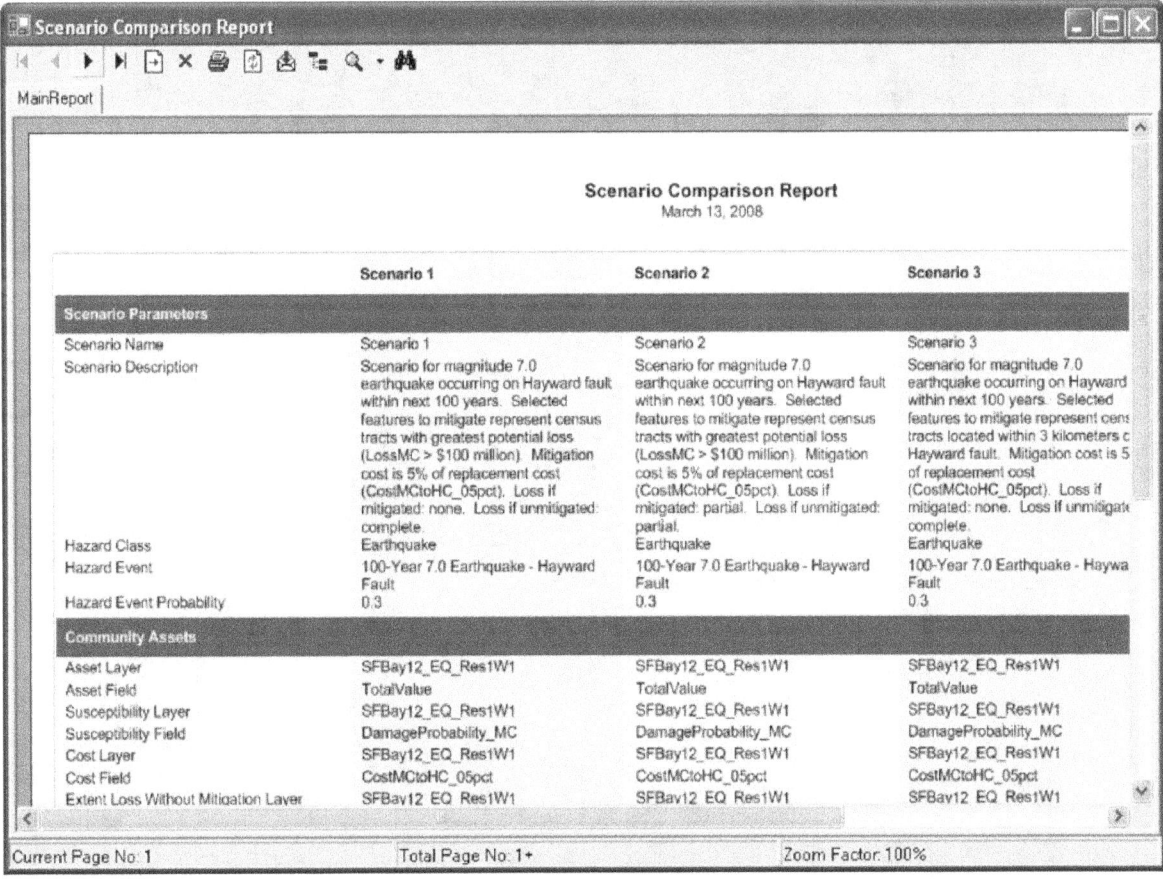

Figure 17. Screen shot showing a sample scenario comparison report shown in the Scenario Comparison Report viewer.

different LUPM operations and execute the LUPM calculations within the ModelBuilder's graphical environment (fig. 18). This gives the analyst the ability to visualize the relationships among the different scenarios. It also gives the analyst the ability to rerun the same basic model with different criteria. These tools are designed to support research activities, rather than production activities, by enabling the analyst to experiment with configuration of the LUPM v1.0 software tools and geoprocessing operations. The LUPM v1.0 Geoprocessing Tools are contained in the USGS LUPM Toolbox.

The analysis (equivalent to a scenario in the PM Tool packages) is the basic unit for the tools in the LUPM v1.0 Geoprocessing Tools package. A model can include and run one or more analyses, each independent of one another. Each scenario is individually represented in a model using a combined sequence of the Make Analysis Layer and the Define Analysis tools.

A basic LUPM v1.0 geoprocessing analysis involves the standard steps described earlier for a scenario—1) identify the data (event, layer, attributes) for a specific analysis layer, 2) integrate one or more analysis layers into an analysis, 3) calculate effects of mitigation, and 4) report and/or save the results (fig. 19). These steps are incorporated into a geoprocessing model that enables the analyst to see and save the steps. The

geoprocessing model serves as the container for the risk-analysis scenario. This allows the analyst to retrieve earlier model runs, much as one would when using the PM Tool's Scenario Manager dialog.

The LUPM v1.0 geoprocessing tools generate XML data strings that are passed to other tools. For example, the Make Analysis Layer tool creates an Analysis Layer XML string containing information about the hazard event, the feature layer (including selected features), and the names of the fields that supply attributes such as asset value, mitigation cost, damage susceptibility, and, optionally, estimates of partial loss. This XML string then serves as input to the Define Analysis tool. The content of these strings should not be modified before passing into another tool, as this could distort the results.

The geoprocessing tools provide a great deal of flexibility in developing an LUPM model in order to support research experimentation. This flexibility includes operations representing data-preparation options that would be difficult to incorporate into the PM User Control and PM ArcGIS Extension tools and the ability to try variations in the model processing. However, this flexibility also creates the potential for errors to enter into the calculations. For example, nothing prevents the value data for a given location from being used more than

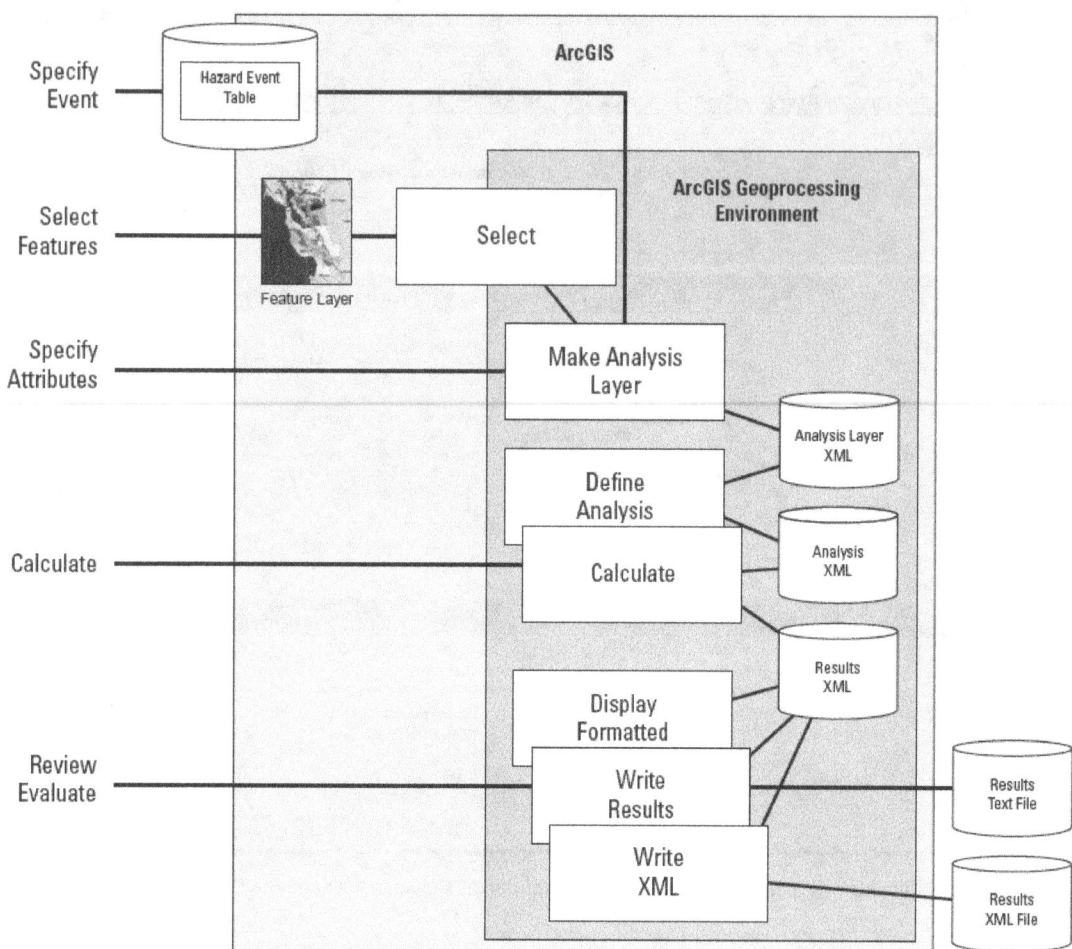

Figure 18. Diagram showing the LUPM v1.0 Geoprocessing Tools activity flow.

once. The LUPM v1.0 Geoprocessing Tools will generate warning messages when these conditions are detected, but the current release of the tools favors flexibility over control.

Create Hazard-Event Data

Hazard-event data for the LUPM v1.0 geoprocessing tools is accessed from a standard ArcGIS data table, such as a .dbf file, a geodatabase table, a tab- or comma-delimited text file, or an Excel spreadsheet. The data may be entered using any standard database or text editing software, as well as Excel.

The table must contain at least three fields. The first field is the event name, and it should be named "Event" or "Name." The second field is the event probability, and it should be named "Probability." The third field is the event description, and it should be named "Description." The table is then added to the ArcMap document, or it may be accessed directly with the geoprocessing tool's file browser.

Select Features to Mitigate

Selecting features for mitigation may be performed in the ArcGIS environment using the Select By Attribute and Select By Location commands. Selection may also be incorporated into the geoprocessing models using the Select By Attribute and Select By Location geoprocessing tools. This ensures that the same feature-selection criteria are applied across multiple model runs.

Create and Execute Scenarios

The first step in creating a scenario using the LUPM v1.0 Geoprocessing Tools package is to prepare the feature layer for analysis using the Make Analysis Layer tool. This tool associates a specific hazard event from the hazard-event table with a feature layer. If more than one record is provided , the tool processes only the first hazard-event record. Specific records may be selected using the standard Make Table View tool.

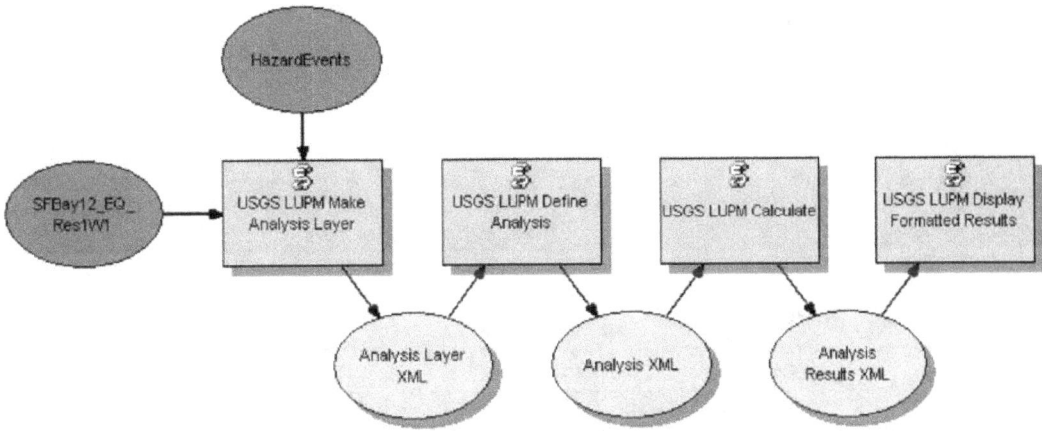

Figure 19. Diagram showing a basic LUPM v1.0 geoprocessing

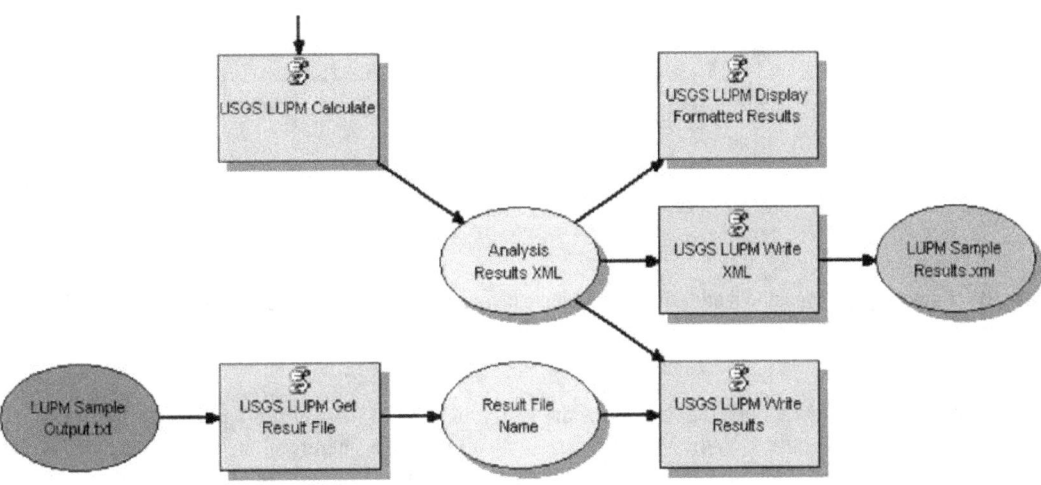

Figure 20. Diagram showing the LUPM v1.0 Geoprocessing Tools output

The Make Analysis Layer tool also identifies the fields containing the feature's attributes (value, damage susceptibility, cost, etc.) corresponding to that particular hazard event. The fields should be appropriate for the specified event (such as susceptibility to earthquake damage for an earthquake event). Finally, the user may choose to enter a string to represent the monetary units used for value and cost (default is "$"). The output from the Make Analysis Layer tool is an XML string containing the hazard information (event name, probability, and description), the feature layer in XML form, and the names of fields containing the asset value, damage susceptibility, mitigation cost, partial losses when mitigated or unmitigated, and the monetary units.

The second step in scenario creation is to use the Define Analysis tool to collect a set of analysis layers into an analysis, or scenario. The tool accepts analysis layer XML strings created by the Make Analysis Layer tool and combines them with a name and description for the analysis. The output from

Define Analysis is an XML string containing the analysis name and description and the XML for each of the analysis layers.

The final step in scenario processing is to use the Calculate tool to run the LUPM calculations based on the defined analysis. The Calculate tool takes the analysis XML and passes that information into the core LUPM libraries. The output from the Calculate tool is an XML string containing the analysis in XML format, the results, and any errors identified during processing.

Evaluate Results

The XML string from the LUPM calculations may be passed into one of several tools for further processing (fig. 20). The Display Formatted tool creates a simple formatted on-screen display of the calculation results in the geoprocessing-status window. The Write to XML File tool writes an XML string to a file. This tool

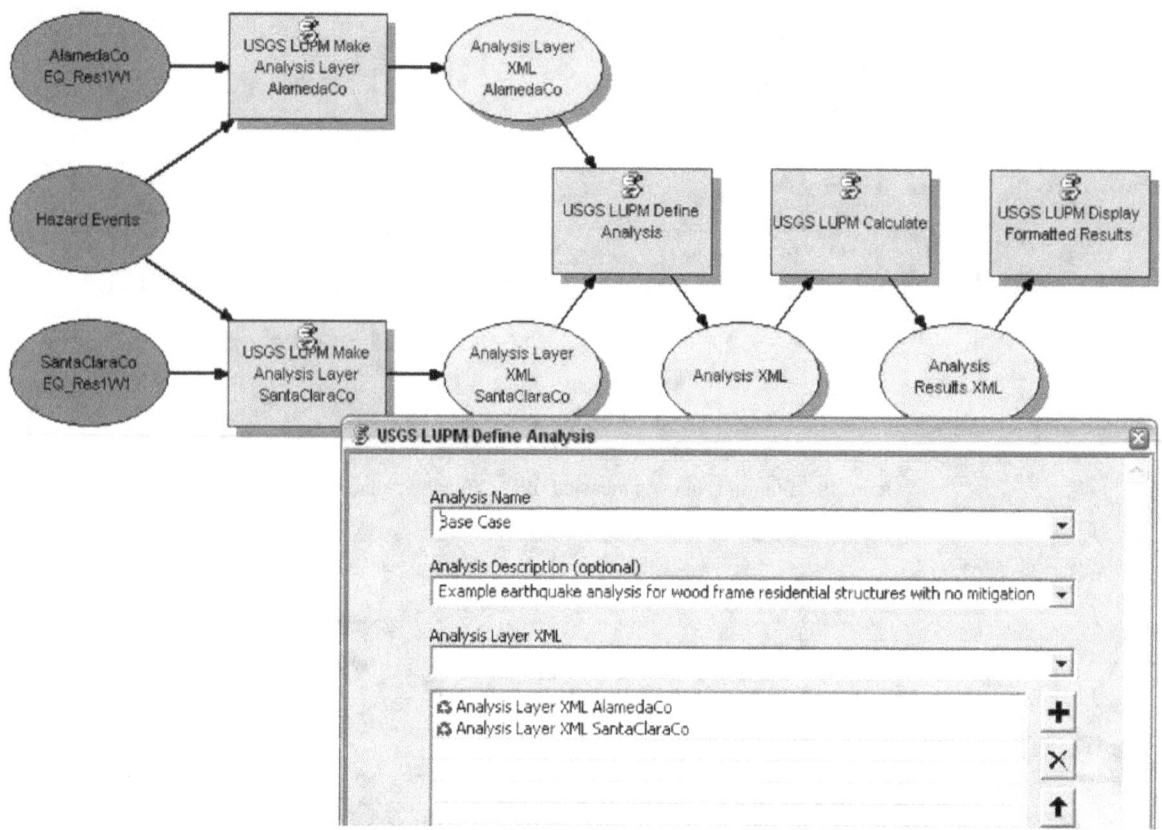

Figure 21. Sample geoprocessing model showing multiple layers in an LUPM v1.0 Geoprocessing Tools analysis.

is typically used with the analysis results, but it may be used with any of the intermediate XML strings created by the Make Analysis Layer or Define Analysis tools.

Finally, the Write Results tool appends the analysis results, along with appropriate field- name column headings, to a tab-delimited file for import into other applications, such as Excel. The path to the file is obtained using the Get Result File tool, which is then passed along as an input to the Write Results tool. This procedure enables appending data to an existing text file.

Additional LUPM v1.0 Geoprocessing Tools Capabilities

An analysis may include multiple analysis layers (fig. 21). This enables the analyst to use data from several feature layers in a single analysis, or scenario, without having to physically merge the layers into a single layer. Each analysis layer (in this example, one for Alameda County and one for Santa Clara County) includes the hazard event, as well as the feature-layer information. The analysis layers are then combined into an analysis using the Define Analysis tool. The analysis is then passed to the Calculate tool. The results are reported as a single analysis for all of the included layers (figs. 21 and 22).

[Note: the PM_Math library currently applies only one event to all features. Therefore, the event associated with the first analysis layer is used with all features in an analysis. This may change in a future release.]

Similarly, multiple analyses may be defined and passed to the Calculate tool (fig. 23). Each analysis may be based on a different selection and/or a different event, but each analysis is calculated independently of other analyses. No information from any analysis is shared with other analyses. The results for each analysis are reported separately (fig. 24).

The geoprocessing model retains the information describing the choices made for a given scenario. These include the choice of hazard event and the fields in the feature layers specifying the asset value, mitigation cost, damage susceptibility, and estimates of partial loss. The model can also be constructed to capture information about the feature-selection criteria used for the scenario, such as the selection of features for mitigation based on attributes or location. Figure 25 shows feature selection using an attribute query. In this case, only census tracts with an estimated loss greater than $10,000,000 are selected for mitigation using the standard ArcGIS Select Layer By Attribute tool.

Figure 26 shows feature selection by proximity to another feature, in this case, census tracts within 3 kilometers of the Hayward Fault. The Hayward Fault features are selected from a layer of quaternary faults (Select Layer By Attribute) and are then used to select the tracts (Select Layer By Location). The LUPM calculation is then completed using the selected tracts.

LUPM Tutorials

The following tutorials will demonstrate how to use the PM Tool-based packages and the LUPM v1.0 Geoprocessing Tools package. Some familiarity with the ArcGIS environment is assumed. The LUPM Version 1.0 software should be installed prior to using any of these tutorials (see appendix A for installation instructions). The software contains the LUPM v1.0 packages, the tutorial sample data, and a sample application using the PM User Control package. Prepare for the tutorial by placing a copy of the Sample Resources folder into your local directory. By default, the Sample Resources folder is installed in the <drive>:\Program Files\USGS\LUPM directory.

The sample data for the tutorial contains data for census tracts prepared using HAZUS-MH (version MR2) software. HAZUS-MH is an ArcGIS tool developed by the Federal Emergency Management Agency (FEMA) that provides a hazard-event scenario-driven approach to estimating losses caused by various types of natural hazards, including earthquakes, floods, and hurricanes. The scenario used for this sample was a probabilistic magnitude 7.0 earthquake event impacting the San Francisco and Monterey Bay Areas. This event was assumed to have a 30 percent probability of occurring within the next 100 years (a 100-year return period). The mitigation objective for this scenario involved upgrading building structures to switch from a moderate building code (MC) to a stricter building code (HC). In an effort to minimize the sample size, only a portion of the output generated from the HAZUS-MH run—data for light wood-frame, single-family residential structures—was selected for inclusion in the sample-data set.

The sample-data set contains a number of fields whose values were calculated from various sources. These fields include property value, damage susceptibility, partial losses, and mitigation cost. Property value was calculated as the average value per home multiplied by the number of homes in the census tract, as specified in the 2000 U.S. Census of Population and Housing. Damage susceptibility was based on the proportion of structures expected to suffer varying degrees of damage based on the five HAZUS damage categories: none, light, moderate, extensive, and complete. LUPM treats damage susceptibility as a probability of suffering damage, but does not differentiate the categories of damage. For the example presented here, damage probabilities were derived from the HAZUS proportions, where the damage probability was equal to the sum of proportions for three of the damage categories: moderate, extensive, or complete damage. Partial-loss values were derived using the original HAZUS damage estimates, recalculated based on the switch from MC to HC (see appendix B for more details on the fields in the sample database and how the values were calculated). Mitigation costs were unknown for upgrading MC structures to meet the stricter HC code. Therefore, costs estimated at 5 percent and 10 percent of replacement cost (the HAZUS exposure variable) are included for this sample.

```
Portfolio Analysis 1 of 1
Name: Base Case
Example earthquake analysis for wood frame residential
    structures with no mitigation
Layers: AlamedaCo_EQ_Res1W1(Polygon)
SantaClaraCo_EQ_Res1W1(Polygon)
Result 1 of 1


Event Name = Sample 7.0 Earthquake
Event Probability = 0.30


Num Locations = 662
Num Locations Mitigated = 0
Original Wealth = $350,095,451,919
Total Asset Value = $350,095,451,919
Total Cost = 0.00
Mitigated Wealth = 0.00
Expected Loss = $27,182,581,987
Expected Loss Variance = $1,742,639,125,398,780,100,000
Expected Loss Std Dev = $41,744,929,338
Expected Wealth = $322,912,869,932
Expected Wealth Variance = $1,742,639,125,398,780,100,000
Expected Wealth Std Dev = $41,744,929,338
Expected Return = 0.00
Expected Return Variance = 0.00
Expected Return Std Dev = 0.00
```

Figure 22. Results from using multiple layers in an LUPM v1.0 Geoprocessing Tools analysis.

PM Tool Tutorial

This tutorial demonstrates how to develop LUPM scenarios using PM Tool-based LUPM v1.0 packages—the PM User Control and PM ArcGIS Extension. The tutorial is divided into three parts. Part 1 covers basic PM Tool operations, including launching the tool and using PM Database Setup, the Hazard Events Manager, and the Scenarios Manager. Part 2 covers creating, editing, and managing LUPM scenarios. Finally, Part 3 covers analyzing scenario results using reporting tools, which are accessed from the Scenarios Manager dialog.

Part 1: Basic PM Tool Operations

Part 1 of this tutorial covers the fundamentals of using the PM Tool, guiding a user through the basic operations of the

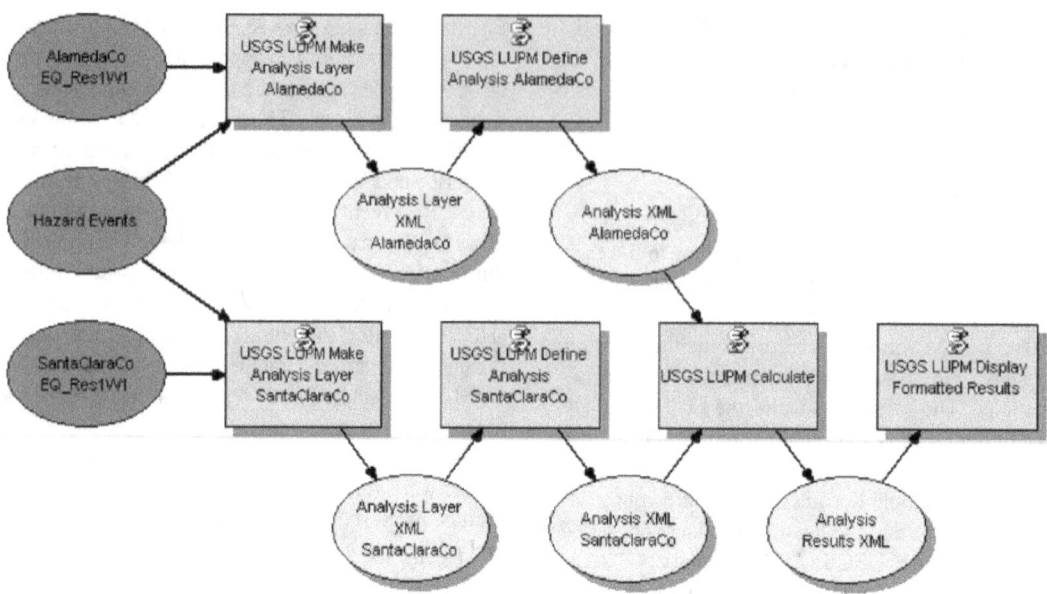

Figure 23. Sample geoprocessing model showing multiple analyses in an LUPM v1.0 Geoprocessing Tools calculation.

tool beginning with lessons on launching the tool and selecting a PM Database. Next, the Hazard Events Manager is discussed along with exercises on using this dialog to create hazard data, including hazard-class and hazard-event data. Finally, the Scenarios Manager is introduced along with an exercise on creating scenario sets that will be used in the second part of this tutorial on developing LUPM scenarios.

Launching the PM Tool

The PM Tool can be launched from the PM User Control using the PM Tool icon located in the control's map-interface toolbar (see fig. 5). A sample application that uses the PM User Control is included in the software and can be used for this tutorial. The application is located in the PM User Control folder, which is in the LUPM v1.0 install directory. The PM Tool can also be launched from ArcMap by using the PM ArcGIS Extension, which provides a PM Tool icon added to the ArcMap application toolbar (see fig. 4). Launching the tool will open the PM Tool's dialog window (see fig. 6), which contains button controls for opening other dialogs, including the PM Database Setup, Hazard Events Manager, and Scenario Manager.

Selecting a PM Database

A PM Database is a custom Microsoft Access database used for storing data created using the PM Tool. PM Databases are managed using the PM Database Setup dialog, accessed from the PM Tool dialog window. This dialog allows you to create a new PM Database or to select an existing database. A PM Database must be selected prior to performing activities using either the Hazard Events Manager or the Scenarios Manager. The following exercises will cover creating and selecting a PM Database using the PM Database Setup dialog.

Creating a PM Database

From the PM Tool dialog window, click Database Setup to open the PM Database Setup dialog window.

Click New to open the Save Microsoft Access File As dialog window.

Specify the location to save the PM Database file to and name the file "myPM.mdb." Click Save to exit. Click OK to the prompt that informs you that the database has been created. The full pathname of the database file should now appear in the PM Database textbox.

Click OK to select the database and exit the PM Setup dialog. The full pathname of the selected database file should now be displayed in the left-hand side of the status bar located at the bottom of the PM Tool dialog window (fig. 27).

Selecting an Existing PM Database

From the PM Tool dialog window, click Database Setup to open the PM Database Setup dialog.

Click Browse to open the Browse Microsoft Access Files dialog window.

Locate and select the "myPM.mdb" file that you created previously. Click Open to exit. The full pathname of the file should now appear in the PM Database textbox.

Click OK to select the database and exit the PM Database Setup dialog.

Managing Hazard Data

The following exercises will cover using the Hazard Events Manager to create, edit, and delete hazard-class and hazard-event data. To begin, click the Hazard Events Manager button from the PM Tool dialog window to open the Hazard Events Manager dialog.

Portfolio Analysis 1 of 2
 Name: Analyze Alameda County
 Example earthquake analysis for wood frame residential
    structures
 Layer: AlamedaCo_EQ_Res1W1(Polygon)
 Result 1 of 1

 Event Name = Sample 7.0 Earthquake
 Event Probability = 0.30

 Num Locations = 321
 Num Locations Mitigated = 0
 Original Wealth = $124,311,882,829
 Total Asset Value = $124,311,882,829
 Total Cost = 0.00
 Mitigated Wealth = 0.00
 Expected Loss = $11,099,434,041
 Expected Loss Variance = $293,171,440,190,369,010,000
 Expected Loss Std Dev = $17,122,249,858
 Expected Wealth = $113,212,448,787
 Expected Wealth Variance = $293,171,440,190,369,010,000
 Expected Wealth Std Dev = $17,122,249,858
 Expected Return = 0.00
 Expected Return Variance = 0.00
 Expected Return Std Dev = 0.00
Portfolio Analysis 2 of 2
 Name: Analyze Santa Clara County
 Example earthquake analysis for wood frame residential
    structures
 Layer: SantaClaraCo_EQ_Res1W1(Polygon)
 Result 1 of 1

 Event Name = Sample 7.0 Earthquake
 Event Probability = 0.30

 Num Locations = 341
 Num Locations Mitigated = 0
 Original Wealth = $225,783,569,090
 Total Asset Value = $225,783,569,090
 Total Cost = 0.00
 Mitigated Wealth = 0.00
 Expected Loss = $16,083,147,945
 Expected Loss Variance = $616,403,099,504,398,960,000
 Expected Loss Std Dev = $24,827,466,635
 Expected Wealth = $209,700,421,145
 Expected Wealth Variance = $616,403,099,504,398,960,000
 Expected Wealth Std Dev = $24,827,466,635
 Expected Return = 0.00
 Expected Return Variance = 0.00
 Expected Return Std Dev = 0.00

Figure 24.

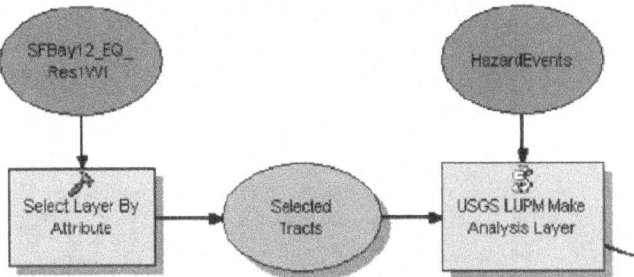

Figure 25. Example showing how to select features by attribute in an LUPM v1.0 Geoprocessing Tools geoprocessing model.

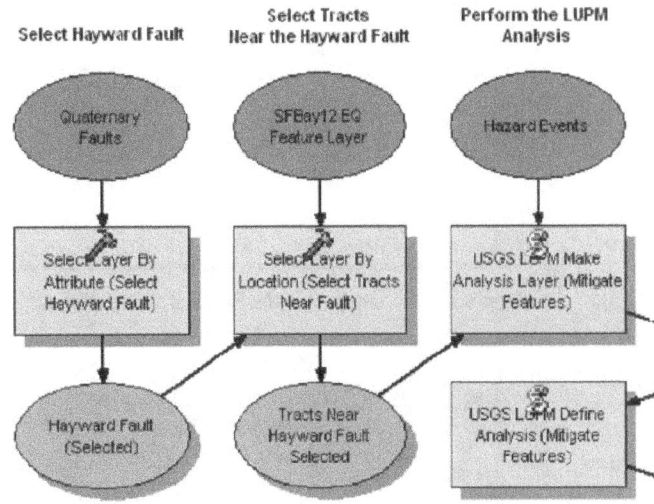

Figure 26. Example showing how to select features by location in an LUPM v1.0 Geoprocessing Tools geoprocessing model.

Adding Hazard Classes

In the Hazard Classes section, click Add to open the Add/Edit Hazard Class dialog window.

Define an "Earthquake" hazard class as depicted in figure 28. Click Save when done.

Repeat steps 1 and 2 to add hazard classes for "Flood," "Landslide," "Hurricane," "Tsunami," and "Volcanic Eruption."

Editing Hazard Classes

In the Hazard Classes section, select the "Earthquake" hazard-class row from the list box and click Edit to display the row in the Add/Edit Hazard Class dialog window.

Edit the Description field, removing the phrase "type of" from the description. Click Save when done.

Deleting Hazard Classes

In the Hazard Classes section, select the "Tsunami" hazard-class row from the list box and click Delete. When prompted to confirm the delete action (fig. 29), click Yes.

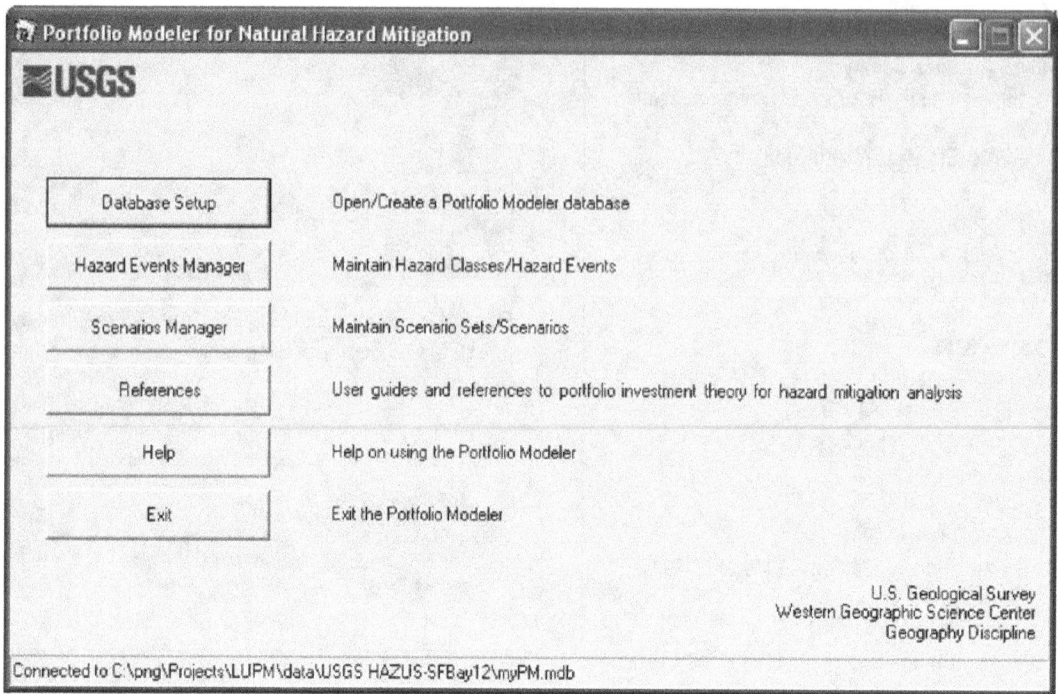

Figure 27.   Screen shot of the PM Tool dialog window: Connection to selected PM Database shown in status bar.

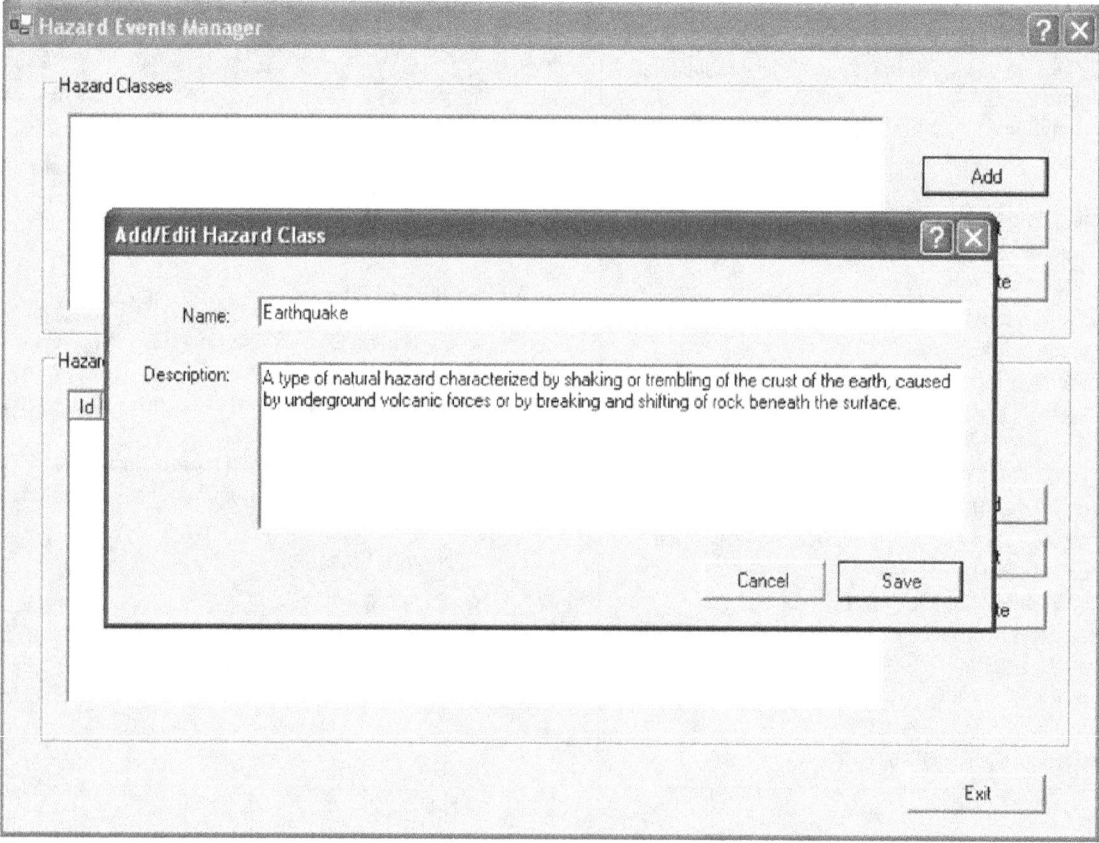

Figure 28.   Screen shot of the Add/Edit Hazard Class dialog window: Input for the "Earthquake" hazard class.

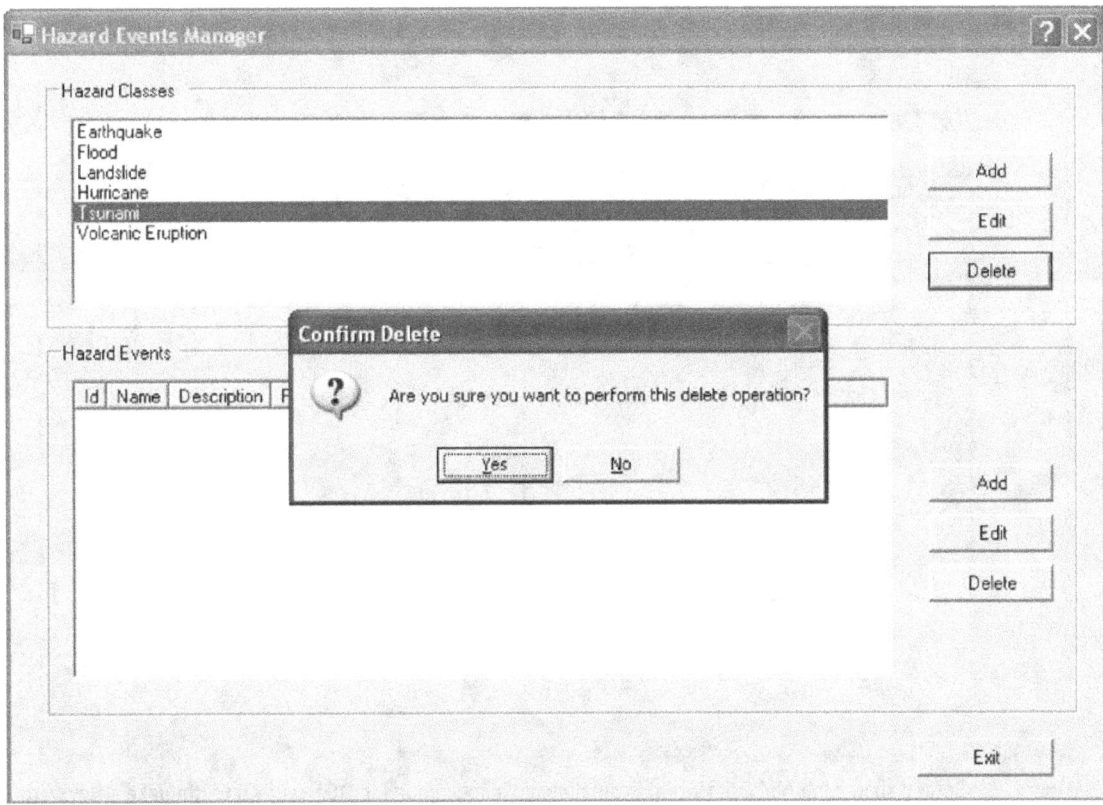

Figure 29. Screen shot of the Confirm Delete dialog window.

Similarly, delete the "Hurricane" and "Volcanic Eruption" hazard-class rows. Hold down the Ctrl key while selecting the desired rows from the list box to select multiple rows.

Adding Hazard Events

In the Hazard Classes section, select the "Earthquake" hazard-class row from the Hazard Classes list box.

Click the Add button located in the Hazard Events section to open the Add/Edit Hazard Event dialog window.

Enter the field values for the hazard event as presented in fig. 30. Click Save when done. The new hazard-event row should now appear in the Hazard Events list box. This event is associated with the currently selected earthquake-hazard class.

Repeat the above steps to add another earthquake-hazard event, setting the field values as follows:

Name:	100-Year 7.0 Earthquake—Calaveras Fault
Description:	Magnitude 7.0 earthquake occurring on the Calaveras fault within the next 100 years
Probability:	0.28

Select the "Landslide" hazard class. Notice that the Hazard Events list box should be empty because this hazard class has no hazard events, yet.

Add an arbitrary landslide-hazard event. You should now see the new landslide-hazard event in the Hazard Events list box.

Click each of the items in the Hazard Classes list box. Notice how the list of items in the Hazard Events list box changes accordingly.

Editing Hazard Events

Select the "Earthquake" hazard-class item from the Hazard Classes list box.

From the Hazard Events section, select the "100-Year 7.0 Earthquake—Hayward Fault" hazard-event item from the Hazard Events list box and click Edit to display the event in the Add/Edit Hazard Event dialog window.

Edit the Probability field, changing its value from "0.25" to "0.30." Click Save when done.

Deleting Hazard Events

Select the "Landslide" hazard class from the Hazard Classes list box.

From the Hazard Events section, select the landslide-hazard event item that you created earlier from the Hazard Events list box and click Delete. Click Yes when prompted to confirm the delete action.

Repeat these steps to delete the "100-Year 7.0 Earthquake—Calaveras Fault" hazard event. The "Earthquake" hazard class should now only have the "100-Year 7.0 Earthquake

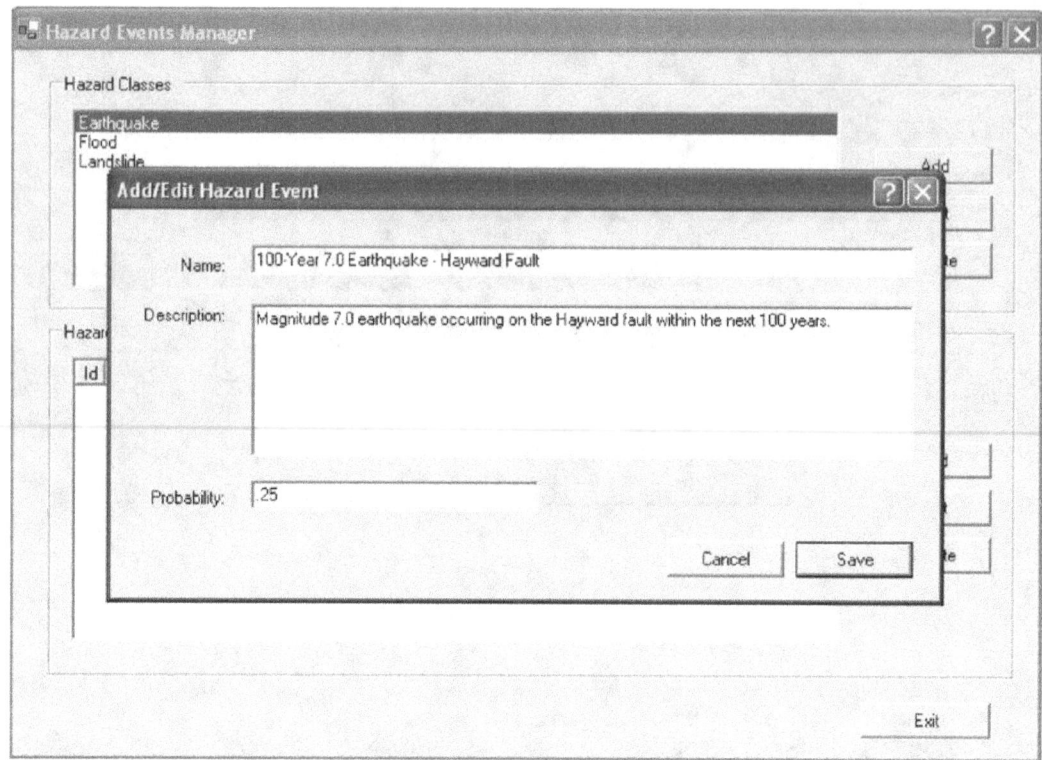

Figure 30. Screen shot of the Add/Edit Hazard Event dialog window: Input for "100-Year 7.0 Earthquake—Hayward Fault" hazard event.

—Hayward Fault" hazard-event item remaining in the Hazard Events list box as illustrated in figure 31.

Click Exit to close the Hazard Events Manager dialog and return control to PM Tool.

Managing Scenario Data

The Scenarios Manager dialog window is used to manage scenario sets and scenario data. The following exercise will cover using the Scenarios Manager to create a scenario set to be associated with scenarios for a hazard event created earlier in the tutorial. Creating, editing, and managing scenarios will be covered at length in the Developing LUPM Scenarios part of this tutorial.

Creating the Scenario Set

From the PM Tool dialog window, click the Scenarios Manager button to open the Scenarios Manager dialog.

You will now create a scenario set to contain scenarios associated with the "100-Year 7.0 Earthquake—Hayward Fault" event. From the Scenario Sets section, click Add to open the Add/Edit Scenario Set dialog window.

Set the values for the Name and Description fields as shown in the figure 32. Click Save when done.

Exit the Scenarios Manager dialog and the PM Tool, returning to the application that is hosting your ArcMap

Document (either the PM User Control sample application or ArcMap). You may exit the application now, or leave it open for the next exercise.

Part 2: Scenario Development

This part of the tutorial demonstrates using the PM Tool to develop eight LUPM scenarios, all of which are for the "100-Year 7.0 Earthquake—Hayward Fault" event. Developing a scenario using the PM Tool is a process that first involves selecting features to mitigate according to an explicit mitigation strategy. This step entails using the query tools, available in ArcMap and the PM User Control's map interface, to select features meeting the selection criteria from the layers contained within an ArcMap document source. The next step involves launching the PM Tool and selecting a PM Database using the PM Database Setup dialog (see part 1 of this tutorial). Finally, a scenario is prepared, run, and saved using the Add/Edit Scenario dialog window accessed from the Scenarios Manager dialog. The following exercises will cover this process for each scenario.

To begin this part of the tutorial, launch the PM User Control sample application or ArcMap, if necessary, and open the sample ArcMap document file named "SF Bay 12 LUPM example.mxd." This file resides in the Sample Resources folder in the LUPM v1.0 install directory. To open this

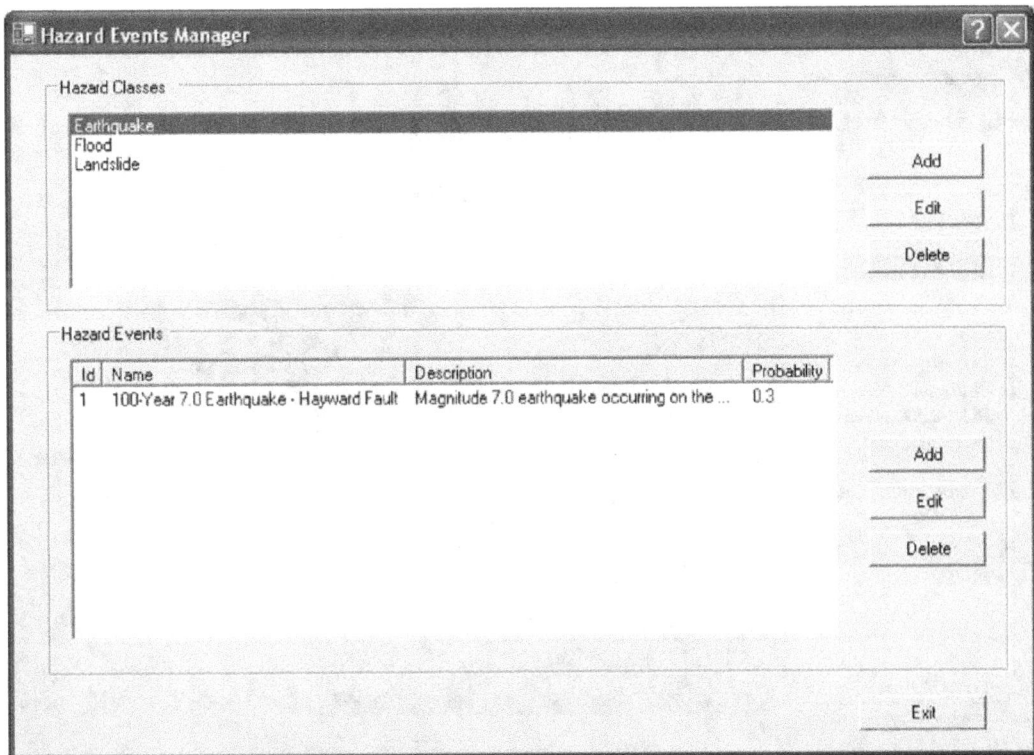

Figure 31. Screen shot of the Hazard Events Manager dialog window showing hazard events associated with the "Earthquake" hazard class.

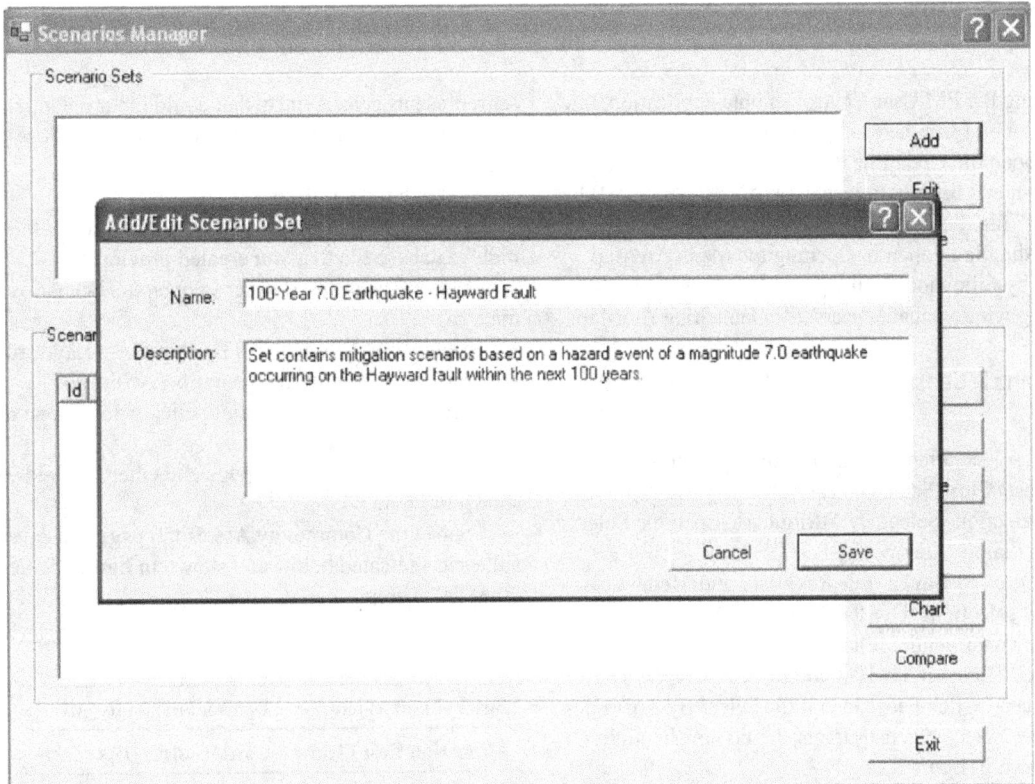

Figure 32. Screen shot of the Add/Edit Scenario Set dialog window: Input for "100-Year Earthquake—Hayward Fault" scenario set.

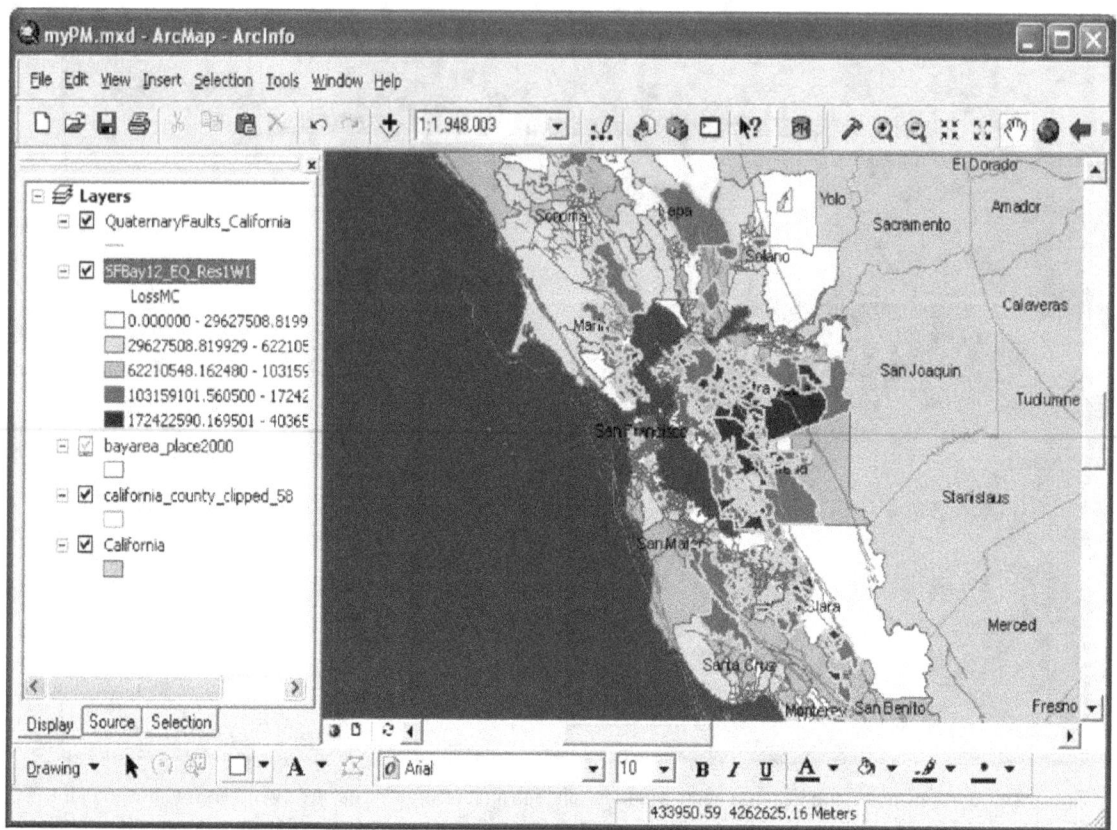

Figure 33. Selected features representing census tracts with the highest potential loss.

document using the PM User Control sample application, click the Open tool icon from the toolbar, or select File>Open from the menu to open the Open File dialog window. Next, use the Look in drop-down list box to locate the "SF Bay 12 LUPM example.mxd" file. Select the file and click Open, or simply double-click the file to open it. Opening the file in ArcMap is performed by following similar procedures or by using the start-up dialog window that appears after launching ArcMap.

Scenarios 1 and 2: Census Tracts with Highest Potential Loss

Clear all selected features in the ArcMap document by selecting Select>Clear Selected Features.

Select Selection>Select By Attributes to open the Select By Attributes dialog window.

Set Layer to "SFBay12_EQ_Res1W1" and Method to "Create a new selection." Use the field attributes and operator controls to construct the following query statement in the query list box: "[LossMC] > 100000000." Click Apply to execute the query. Click Close to exit the Select By Attributes dialog window. Selected census tracts should now be highlighted as shown in figure 33.

Open the attribute table for the "SFBay12_EQ_Res1W1" layer to verify that 281 out of 1,549 rows were selected. This

set represents census tracts that could incur potential losses greater than $100 million, if left unmitigated. Close the table when done.

Launch the PM Tool.

Use the PM Database Setup dialog to select the "myPM. mdb" database file that you created previously.

Click Scenarios Manager to open the Scenarios Manager dialog.

Select the "100-Year 7.0 Earthquake—Hayward Fault" scenario set item from the Scenario Sets list box.

From the Scenarios section, click Add to open the Add/ Edit Scenario dialog window.

Select the General tab page and enter the field values as shown in figure 34.

Select the Community Assets tab page and enter the field values as indicated below and shown in figure 35. Leave the other fields empty.

Assets Field:	TotalValue
Susceptibility Field:	DamageProbability_MC
Mitigation Cost Field:	CostMCtoHC_05pct

Click Run Scenario to run the scenario through the LUPM calculations. After the scenario run has completed, a

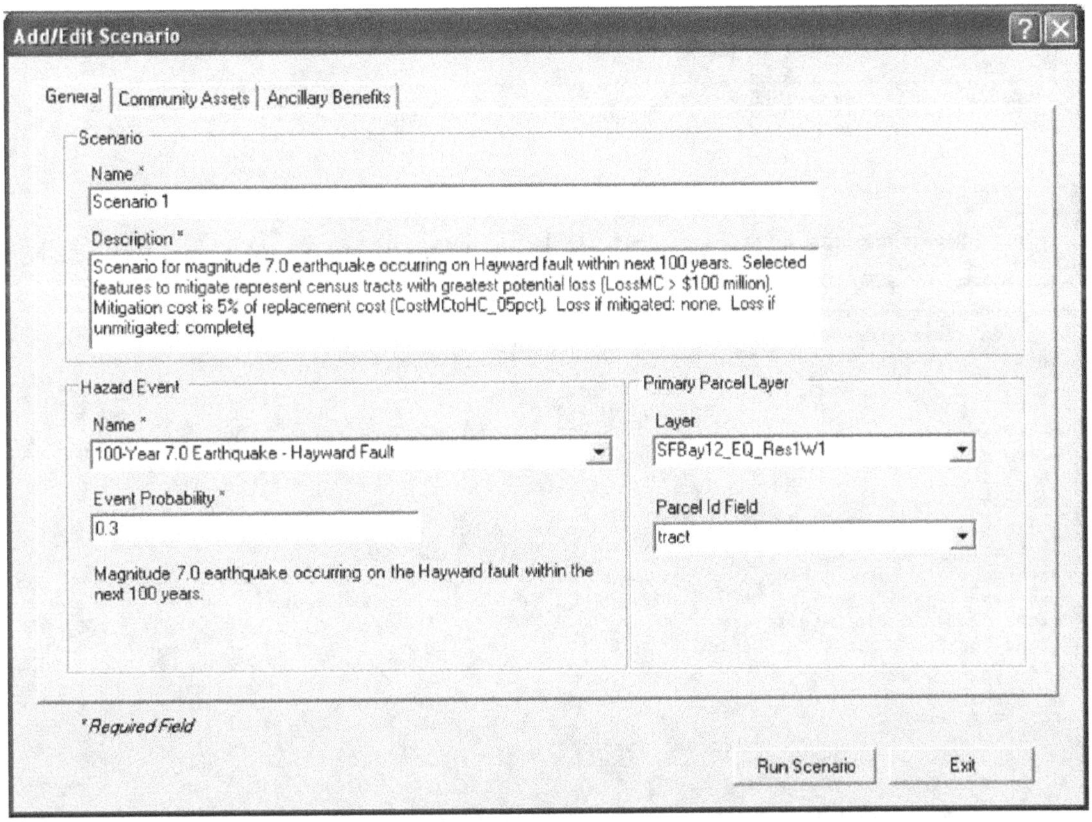

Figure 34. Screen shot of the Add/Edit Scenario dialog window: General tab page input for Scenario 1.

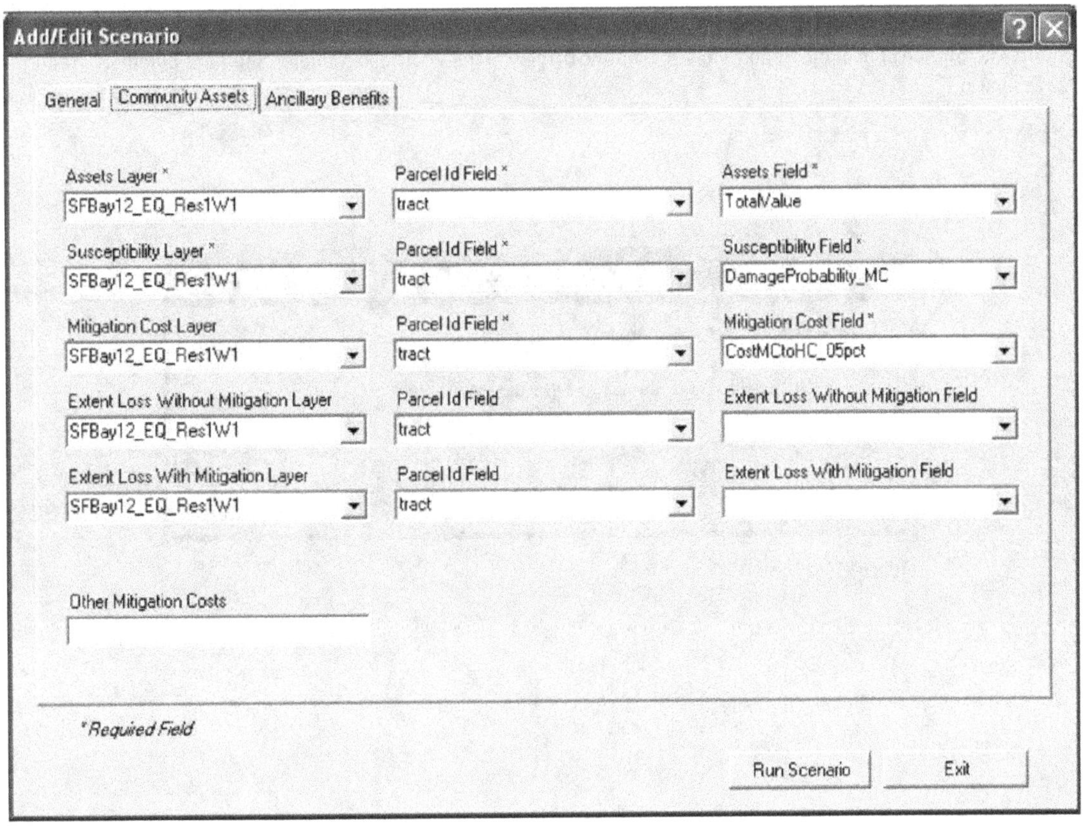

Figure 35.

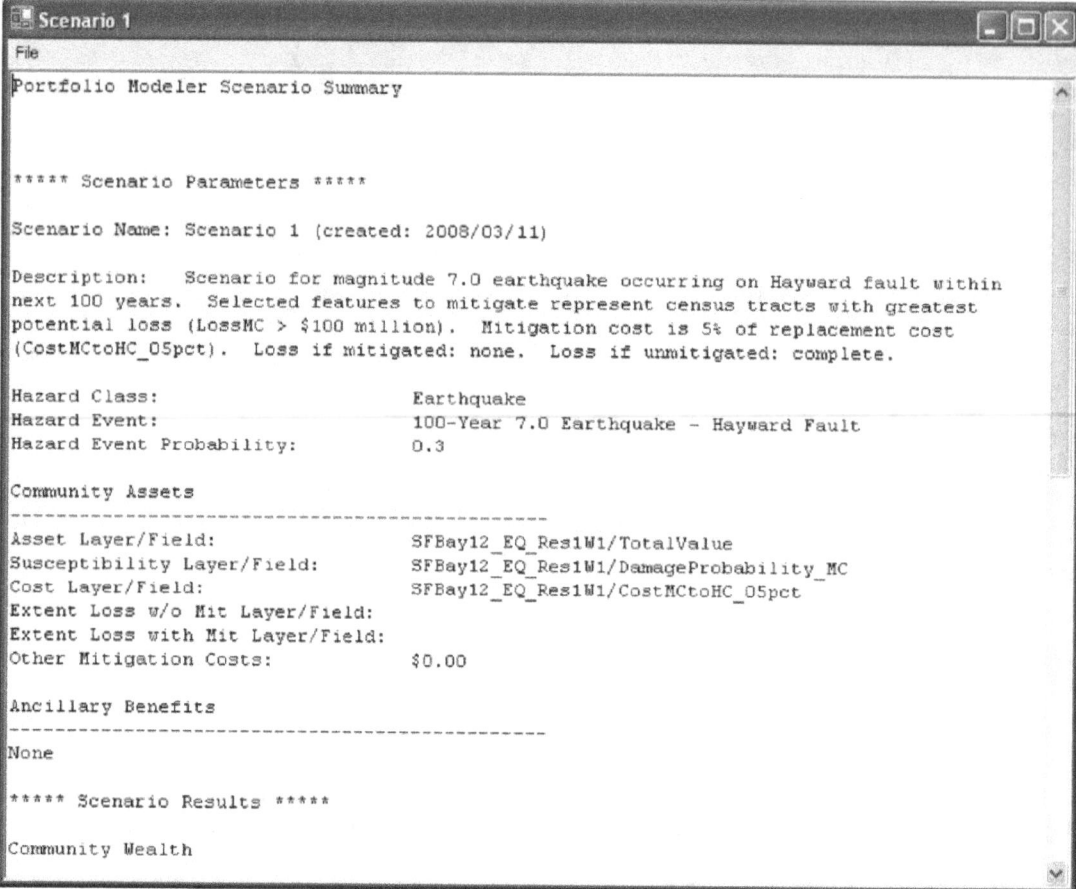

Figure 36. Screen shot of the Report Viewer window: Display of the "Portfolio Modeler Scenario Summary" report for Scenario 1.

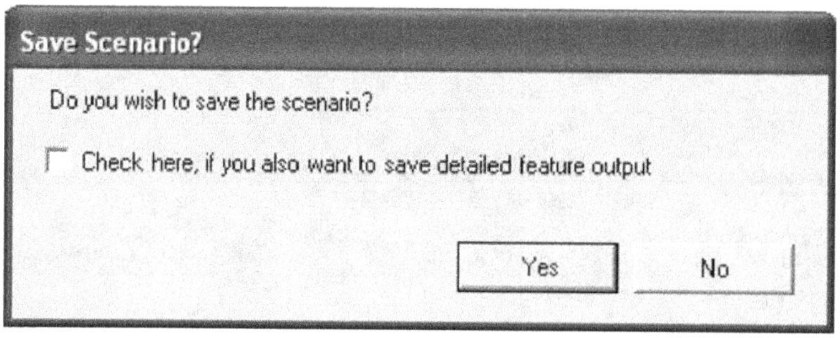

Figure 37.

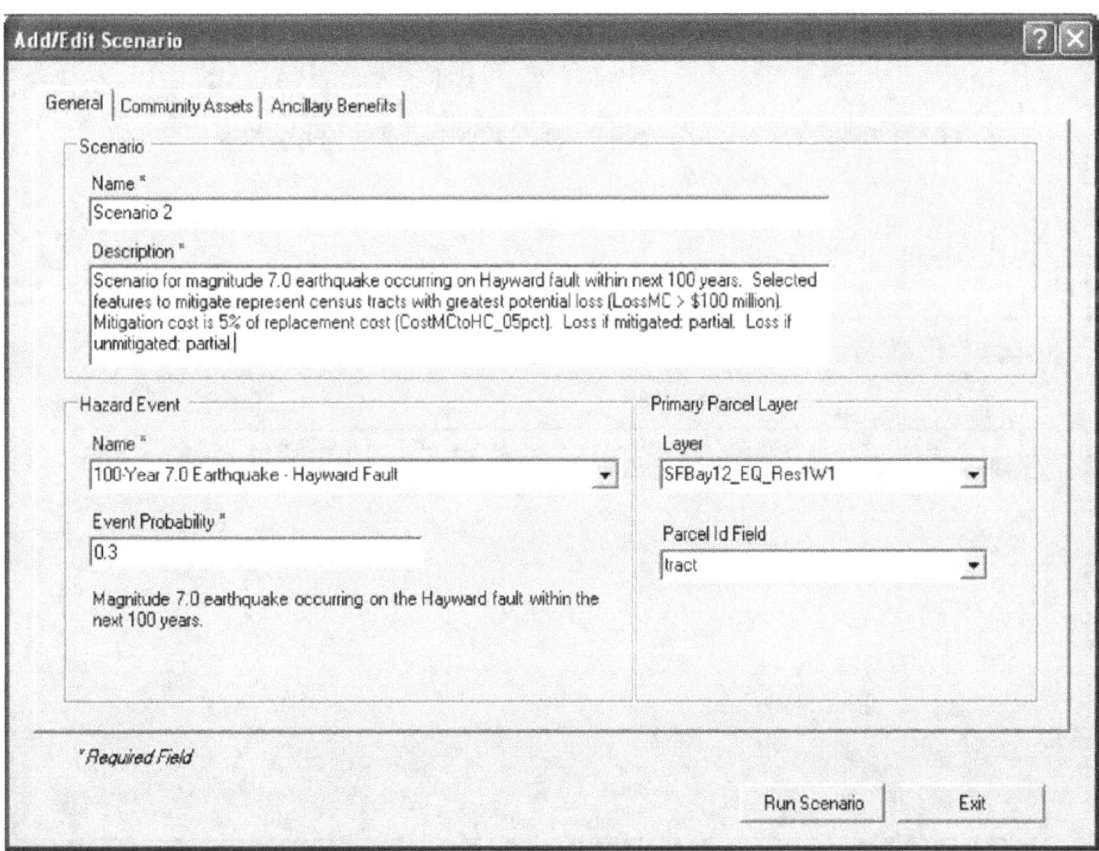

Figure 38. Screen shot of the Add/Edit Scenario dialog window: General tab page input for Scenario 2.

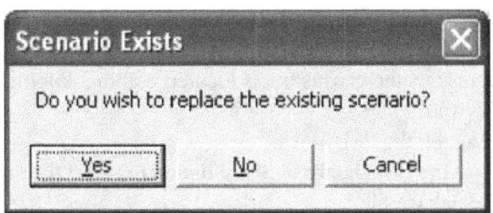

Figure 39. Screen shot of the Scenario Exists dialog window.

Portfolio Modeler Scenario Summary report is generated and displayed in the Report Viewer window as illustrated below (fig. 36). Review the report, and then exit the viewer.

Leave the checkbox that asks if you wish to also save the detailed feature output when saving the scenario unchecked when prompted to save the scenario (fig. 37). Click Yes to save the scenario ("Scenario 1") and exit the prompt window, returning to the Add/Edit Scenario dialog window.

Now modify the previous scenario ("Scenario 1") to include partial-loss estimate data. Select the General tab page and edit the Name and Description fields as illustrated in figure 38.

Select the Community Assets tab page and set the Extent Loss Without Mitigation Field and Extent Loss With Mitigation Field fields as follows:

Extent Loss Without Mitigation Field:	UnmitigatedLoss_Prop
Extent Loss With Mitigation Field:	MitigatedLoss_Prop

Click Run Scenario to run the edited scenario through the LUPM calculations. Review the generated Portfolio Modeler Scenario Summary report and exit the viewer when done.

Click Yes when prompted to save the scenario. The next prompt will ask if you wish to replace the existing scenario with the edited scenario (fig. 39). Click No to save the edited scenario as new ("Scenario 2").

Exit the Add/Edit Scenario dialog window to return to the Scenarios Manager dialog. Notice that the two scenarios you created now appear in the Scenarios list box as shown below (fig. 40). These scenarios are associated with the 100-Year 7.0 Earthquake—Hayward Fault scenario-set item selected in the Scenario Sets list box.

Exit the Scenarios Manager dialog and the PM Tool, returning to the hosting application. You may exit the application now, or leave it open for the next exercise.

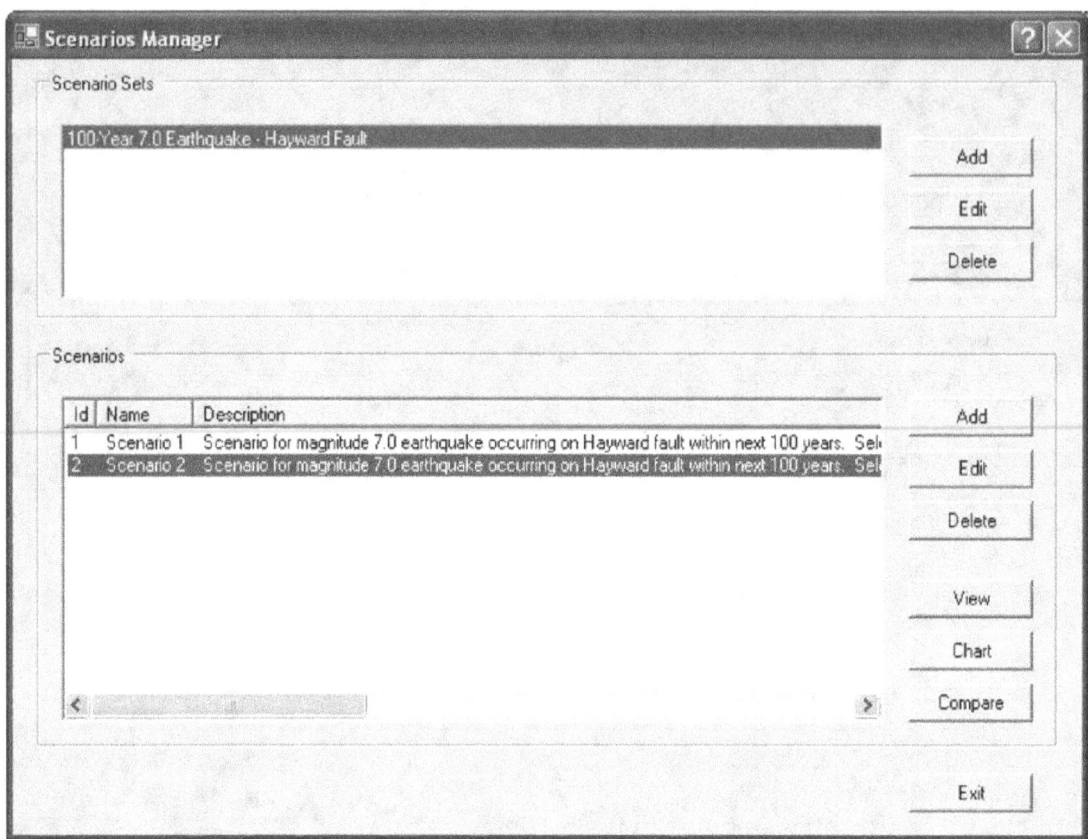

Figure 40. Screen shot of the scenarios listed for "100-Year 7.0 Earthquake—-Hayward Fault" scenario set.

Scenarios 3 and 4: Census Tracts Located Within 3 Kilometers of Hayward Fault

Return to your ArcMap document.

Clear all selected features by selecting Selection>Clear Selected Features from the menu.

Select Selection>Select By Attributes from the menu to open the Select By Attributes dialog window.

Set Layer to "QuaternaryFaults" and Method to "Create a new selection." Use the field attributes and operator controls to construct the following query statement in the query list box: "[Name] LIKE '*Hayward*'." Click Apply to execute the query and exit the window.

Open the attribute table for the "QuaternaryFaults" layer to verify that 100 out of 17,449 rows were selected. Collectively, these rows make up the Hayward Fault.

The second step of this query will determine the features that are located within 3 kilometers of the Hayward Fault. Select Selection>Select By Location from the menu to open the Select By Location dialog window.

Set the fields in this window as shown below (fig. 41).

When finished, click Apply to execute the query, and then close the window. Selected features should now be highlighted as shown in figure 42.

Open the attribute table for the "SFBay12_EQ_Res1W1" layer to verify that 264 out of 1,549 rows were selected. This set represents the census tracts located within 3 kilometers of the Hayward Fault.

Launch the PM Tool.

Use the PM Database Setup dialog to select the "myPM.mdb" database file.

Click Scenarios Manager to open the Scenarios Manager dialog.

Select the "100-Year 7.0 Earthquake—Hayward Fault" scenario set from the Scenario Sets list box.

From the Scenarios section, click Add to open the Add/Edit Scenario dialog window.

Select the General tab page and enter the field values as shown in figure 43.

Select the Community Assets tab page and set the assets, susceptibility, and mitigation cost fields as follows:

Assets Field:	TotalValue
Susceptibility Field:	DamageProbability_MC
Mitigation Cost Field:	CostMCtoHC_05pct

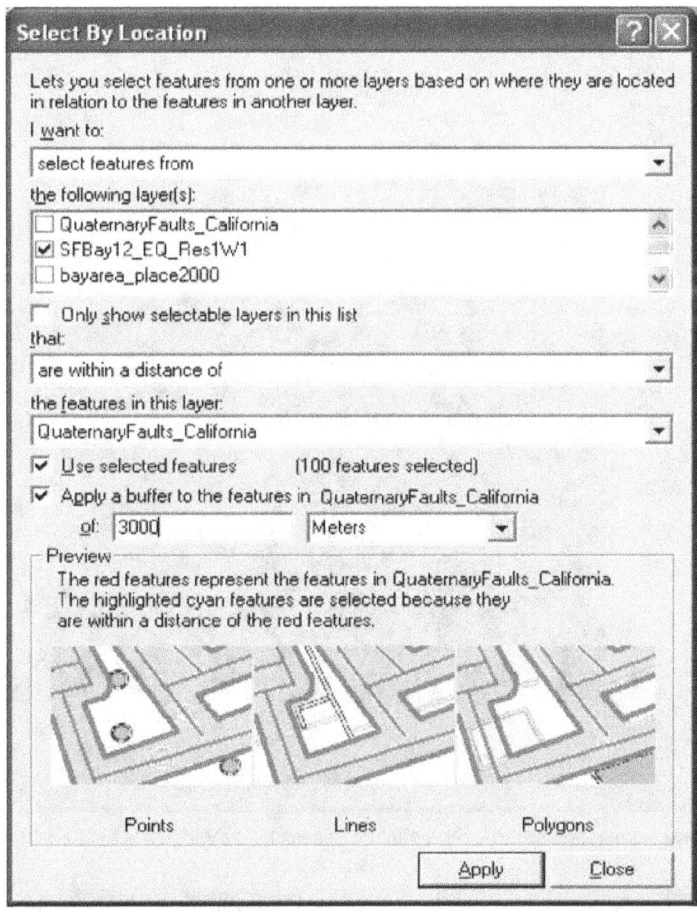

Figure 41. Screen shot showing the Select By Location dialog window: Input for Scenarios 3 and 4.

Click Run Scenario. Review the generated Portfolio Modeler Scenario Summary report and exit the viewer when done.

Save the scenario when prompted to do so (Scenario 3).

Create a modified version of the scenario using partial-loss estimate data. You can edit the scenario while you are still in the Add/Edit Scenario dialog window (as you did before), or you can return to editing the scenario from the Scenarios Manager dialog. You will use the latter option for this exercise. Close the Add/Edit Scenario dialog window, returning you to the Scenarios Manager.

From the Scenarios section, select the previously created scenario ("Scenario 3") from the Scenarios list box. Click Edit to display the scenario in the Add/Edit Scenario dialog window.

Select the General tab page and edit the Name field to reflect that this will be "Scenario 4." Also, edit the text in the Description field to indicate that this scenario will use partial-loss estimate data.

Select the Community Assets tab page and set the Extent Loss Without Mitigation Field and Extent Loss With Mitigation Field fields as follows:

Extent Loss Without Mitigation Field:	UnmitigatedLoss_Prop
Extent Loss With Mitigation Field:	MitigatedLoss_Prop

Click Run Scenario. Review the generated Portfolio Modeler Scenario Summary report and exit the viewer when done.

Click Yes when prompted to save the scenario. Another prompt window will appear and ask if you wish to replace the existing scenario with the edited scenario. Click No to save the edited scenario as new ("Scenario 4"). Exit the Add/Edit Scenario dialog window to return to the Scenarios Manager dialog.

Exit the Scenarios Manager dialog and the PM Tool.

Scenarios 5 and 6: Census Tracts with Greatest Loss and Near the Hayward Fault

In ArcMap or the PM User Control, clear all selected features by selecting Selection>Clear Selected Features from the menu.

Select Selection>Select By Attributes from the menu to open the Select By Attributes dialog window.

Set Layer to "QuaternaryFaults" and Method to "Create a new selection." Use the field attributes and operator controls

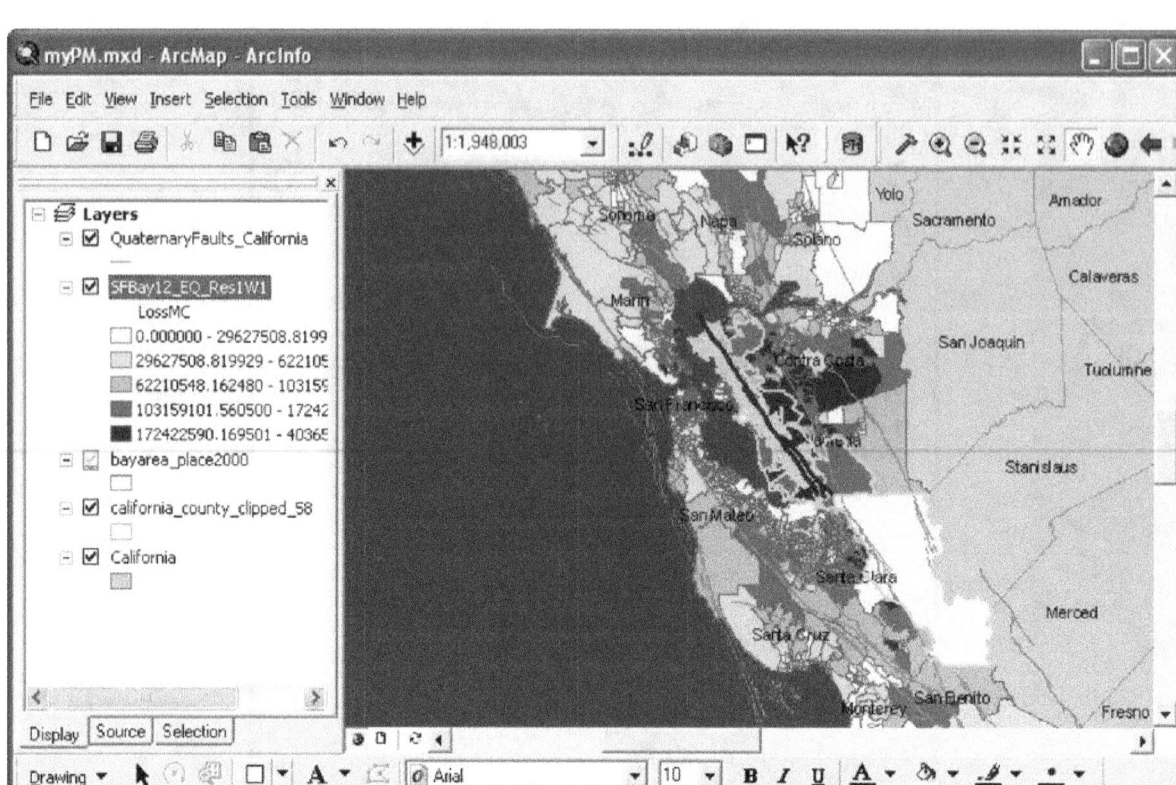

Figure 42. Screen shot showing selected features representing census tracts within 3 kilometers of Hayward Fault.

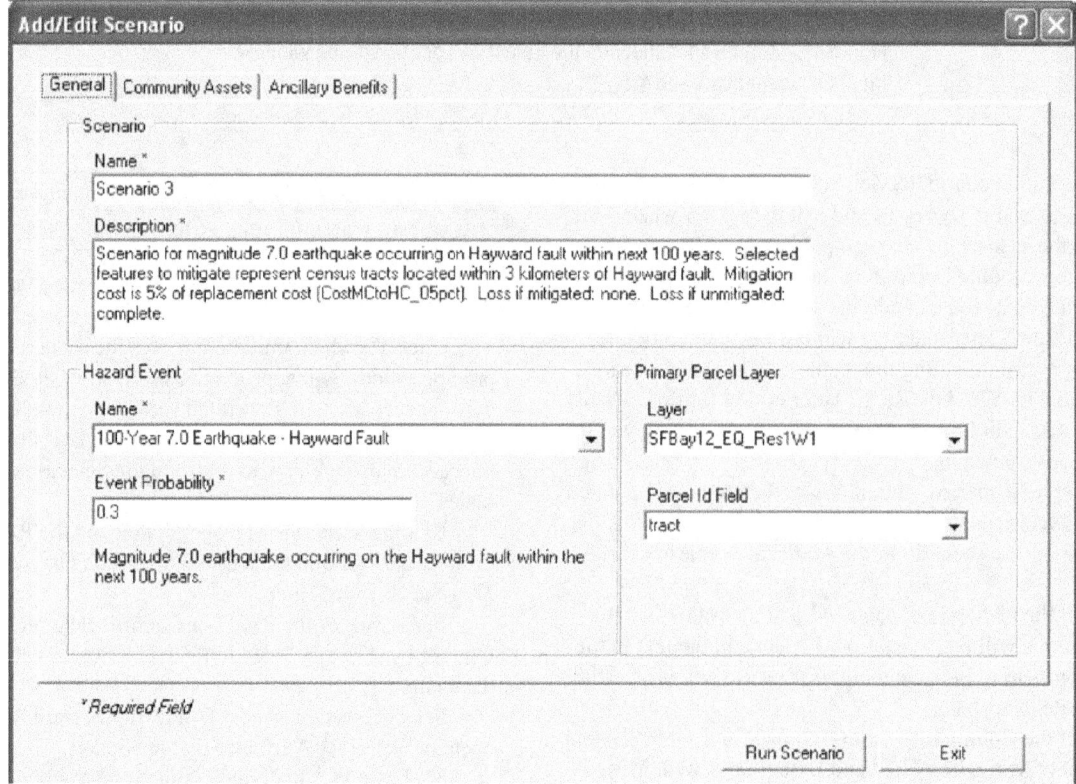

Figure 43. Screen shot of the Add/Edit Scenario dialog window: General tab page input for Scenario 3.

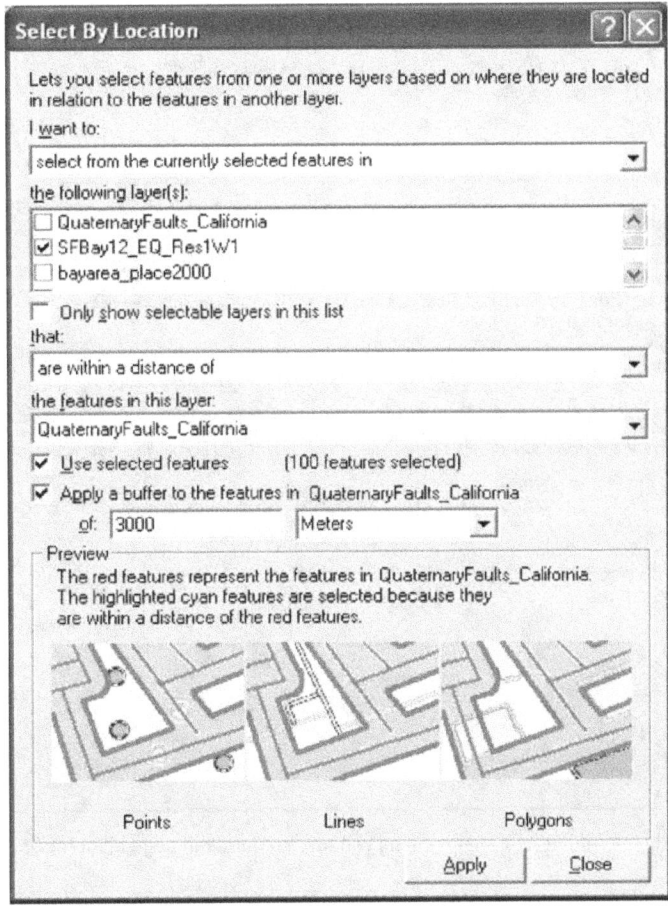

Figure 44. Screen shot showing the Select by Location dialog window: Input for Scenarios 5 and 6.

to construct the following query statement in the query list box: "[Name] LIKE '*Hayward*'." Click Apply to execute the query. Leave the Select By Attributes dialog window opened.

Set Layer to "SFBay12_EQ_Res1W1" and Method to "Add to current selection." Construct the following query statement in the query list box: "[LossMC] > 100000000." Click Apply to execute the query, and then close the window.

Select Selection>Select By Location from the menu to open the Select By Location dialog window.

Set the fields in this window as shown below (fig. 44). Make sure to use "Select from the currently selected features in" item from the drop-down list as shown. This will use the features you have already selected as the basis for the new selection. The resulting selection represents high-potential loss census tracts that are located within 3 kilometers of the Hayward Fault.

When finished, click Apply to execute the query, and then close the window.

Open the attribute table for the "SFBay12_EQ_Res1W1" layer to verify that 65 out of 1,549 rows were selected. This set represents the census tracts with the greatest potential loss that are located within 3 kilometers of the Hayward fault.

Launch the PM Tool and select the "myPM.mdb" database file.

Open the Scenarios Manager dialog and select the "100-Year 7.0 Earthquake—Hayward Fault" scenario set.

From the Scenarios section, click Add to open the Add/Edit Scenario dialog window.

Select the General tab page and enter the field values as shown in figure 45.

Select the Community Assets tab page and set the assets, susceptibility, and mitigation cost fields as follows:

Assets Field:	TotalValue
Susceptibility Field:	DamageProbability_MC
Mitigation Cost Field:	CostMCtoHC_05pct

Click Run Scenario. Review the generated Portfolio Modeler Scenario Summary report, exit the viewer when done, and save the scenario ("Scenario 5").

Create a modified version of this scenario using partial loss estimate data, setting the Name, Description, Extent Loss

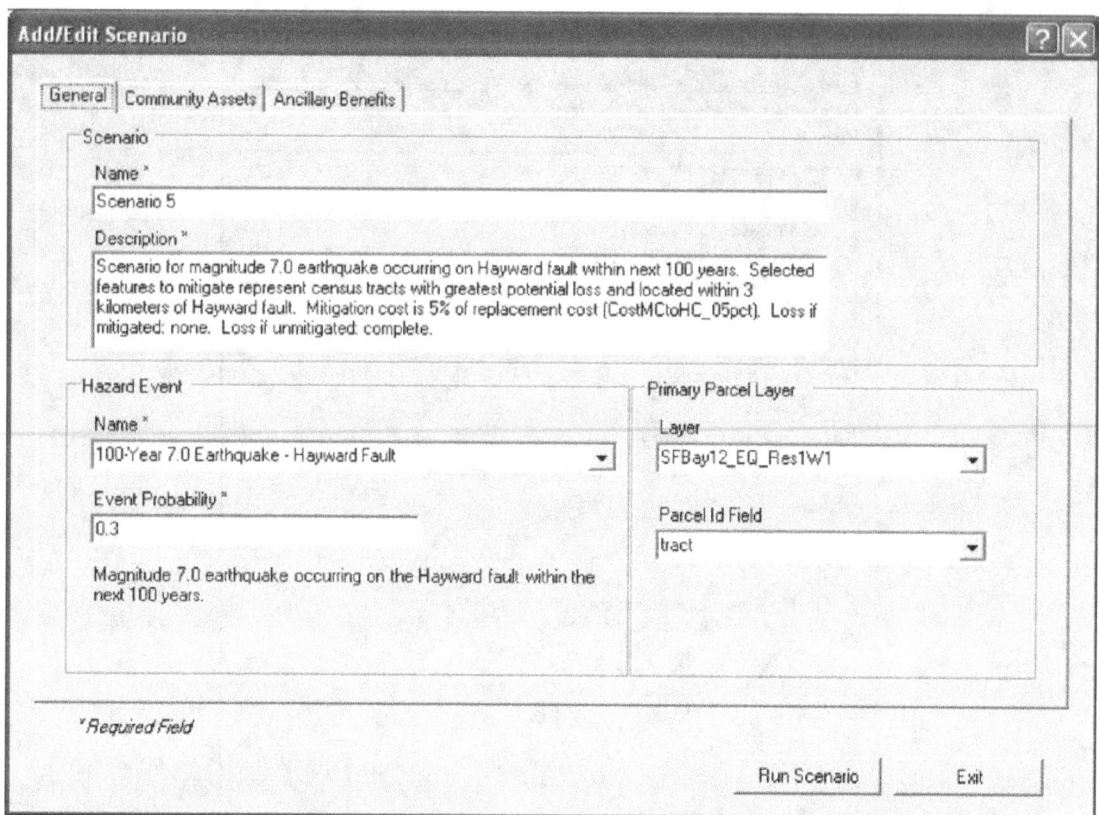

Figure 45. Screen shot of the Add/Edit Scenario dialog window: General tab page input for Scenario 5.

Without Mitigation Field and Extent Loss With Mitigation Field fields appropriately.

Run the scenario, review the output, and save the scenario as new ("Scenario 6"). Exit the Add/Edit Scenario dialog window to return to the Scenarios Manager dialog.

Scenarios 7 and 8: Mitigation Cost Set at 10 percent of Replacement Cost

All the scenarios developed so far have used a mitigation cost field that represents the cost of upgrading a building structure from a moderate code to a higher code as 5 percent of the replacement cost of the structure. The following steps show how to modify the last two scenarios using a mitigation-cost field representing 10 percent of replacement cost.

Select the "100-Year 7.0 Earthquake—Hayward Fault" scenario set. Next, select "Scenario 5" from the Scenarios section and click Edit to display the scenario in the Add/Edit Scenario dialog window.

Select the General tab page. Edit the Name field to reflect that this scenario will become "Scenario 7" and the Description field to indicate that the scenario will assume the mitigation cost to be 10 percent of replacement cost.

Set Mitigation Cost Field to "CostMCtoHC_10pct" in the Community Assets tab page.

Run the scenario, review the output, and save the scenario as new ("Scenario 7"). Exit the Add/Edit Scenario dialog window to return to the Scenarios Manager dialog.

Select "Scenario 6" from the Scenarios section and click Edit to display the scenario in the Add/Edit Scenario dialog window.

Edit the Name field to reflect that this scenario will become "Scenario 8" and the Description field to indicate that the scenario will assume the mitigation cost to be 10 percent replacement cost.

Set Mitigation Cost Field to "CostMCtoHC_10pct" in the Community Assets tab page.

Run the scenario, review the output, and save the scenario as new ("Scenario 8").

Exit the Scenarios Manager dialog and the PM Tool, returning to the hosting application.

Part 3: Analyzing Scenario Results

The third and final part of this tutorial covers several tools used for viewing or analyzing the results of one or more LUPM scenarios. These tools, which are accessed from the Scenarios Manager, include the Report Viewer, the Chart tool,

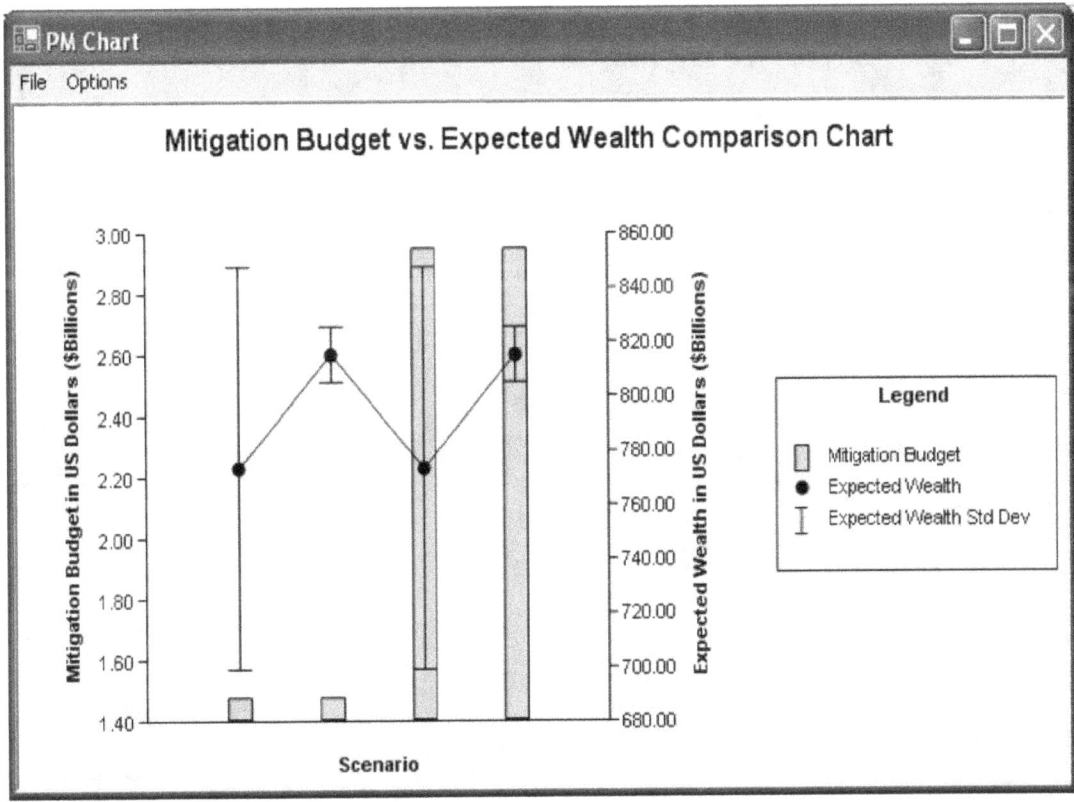

Figure 46. Screen shot of a multiseries standard-error chart.

and the Comparison tool. The following exercises will demonstrate how these tools are used.

To begin, launch the PM Tool from the hosting application. Next, use the PM Database Setup dialog to select the "myPM.mdb" database file. Finally, open the Scenarios Manager dialog to prepare for the exercises to follow.

Using the Report Viewer

Select the "100-Year 7.0 Earthquake—Hayward Fault" scenario set.

From the Scenarios section, select any scenario item from the Scenarios list box and click View to display the report for the scenario in the Report Viewer. Note that the Report Viewer referred to here is the same viewer used to display the output following a scenario run. When used to view a report, as in the case here, the Report Viewer will not prompt to save the scenario when exiting.

The viewer has options for printing the report, as well as saving it as a plain text file. Save the report as a text file by selecting File>Save from the menu to open the Save File As dialog window. Specify a location and filename for the file and click Save when done.

Exit the viewer to return to the Scenarios Manager dialog.

Using the Chart Tool to create a default chart based on Scenarios 5 through 8.

From the Scenarios section, select "Scenario 5" from the Scenarios list box. Next, hold down the Shift key and select "Scenario 8." The range of items between "Scenario 5" and "Scenario 8," inclusive, should now be selected.

From the Scenarios section, click Chart to open the Chart Options dialog.

Click Use Defaults to use default settings for the Chart tool. Click OK to generate and display the chart in the Chart Viewer as illustrated below (fig. 46). Notice that the chart created is a multi-series standard-error chart comparing "Mitigation Budget" to "Expected Wealth" for each of the selected scenarios.

Exit the Chart Viewer to return to the Scenarios Manager.

Now, modify the chart to compare "Avoided Losses" to "Expected Wealth." Click Chart to re-open the Chart Options dialog with Scenarios 5 through 8 still selected.

Select the Data tab page and set the fields according to figure 47.

Select the Text tab page and set the Chart Title and Y Axis Label (L) as follows:

Chart Title:	Avoided Losses vs. Expected Wealth Comparison Chart
Y Axis Label (L)	Avoided Losses With Mitigation in US Dollars

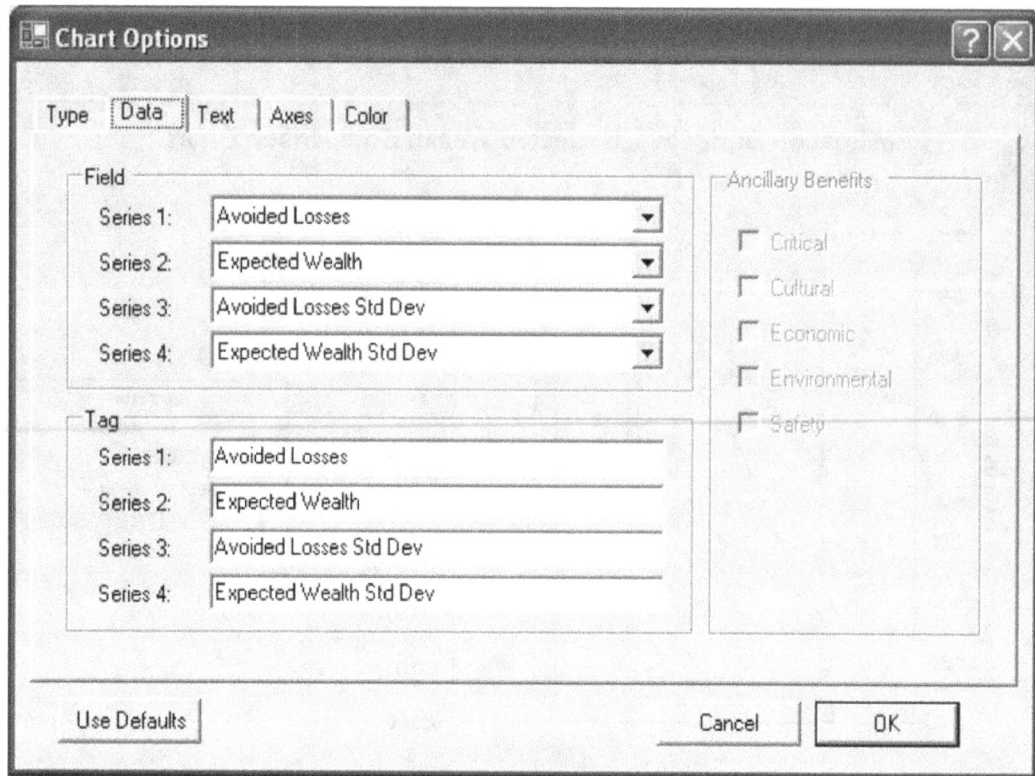

Figure 47. Screen shot showing the Chart Options dialog window: Data tab page input settings.

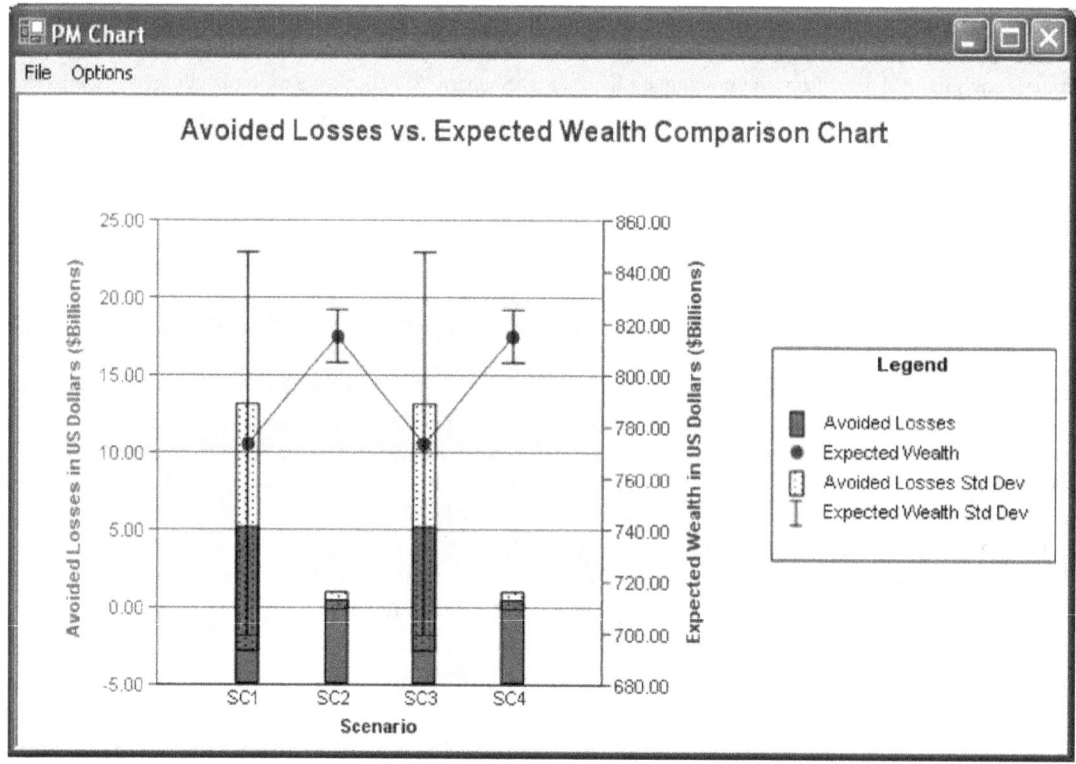

Figure 48. Screen shot of a customized multiseries standard error chart.

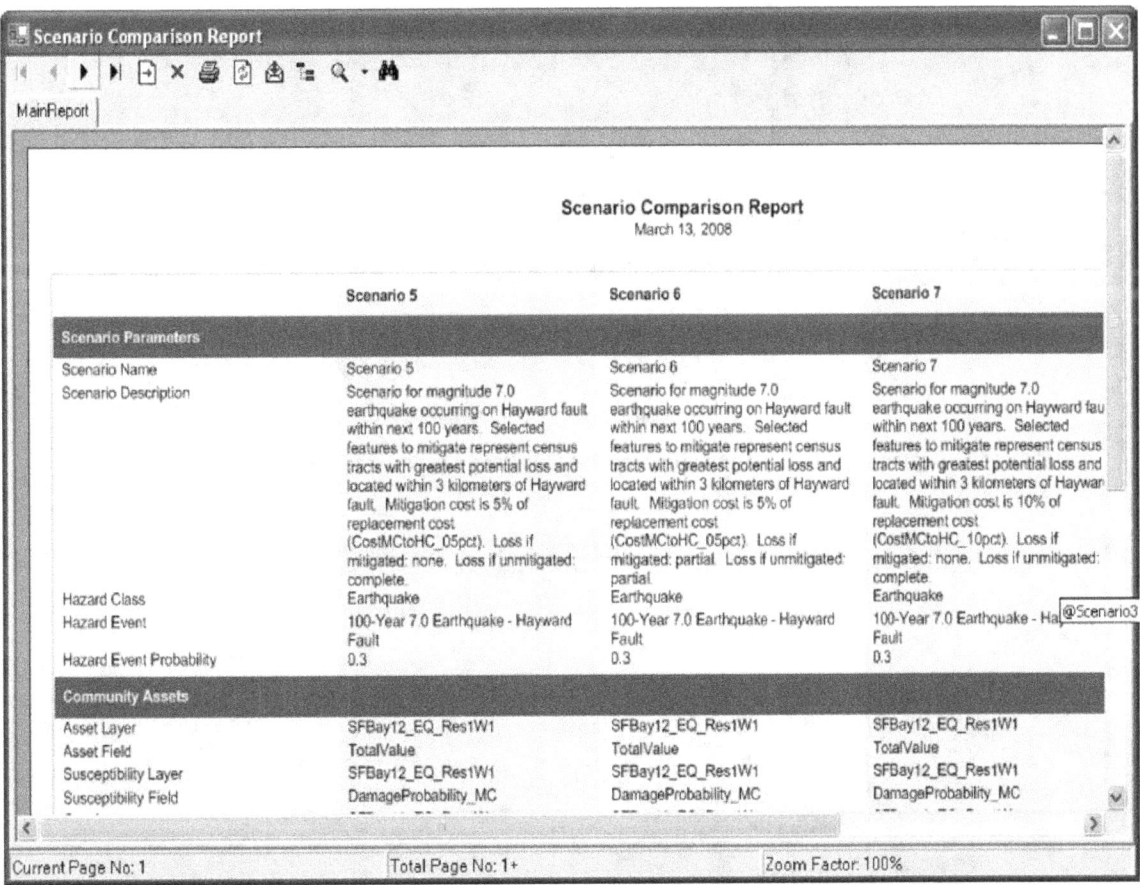

Figure 49. Screen shot of the Scenario Comparison Report Viewer window: Comparison report of Scenarios 5 through 8.

Click OK to generate and view the chart in the Chart Viewer. Exit the viewer when done.

Now, refine the previous chart for Scenarios 5 through 8. Click Chart to reopen the Chart Options dialog.

Select the Axes tab page. This page provides fields used to set the number of step increments to display along both Y-axes, enable the display of tags identifying each scenario, and enable the display of horizontal gridlines. Edit the settings in this page as follows:

Set the Min Steps Y Left field	6
Set the Min Steps Y Right field	6
Show Tags checkbox	Checked
Horizontal (L) checkbox	Checked

Select the Color tab page and select a different color to use for one or more of the chart elements. Click on the color box next to the element to open a color palette window to set the element's color. Choose a color and click OK to apply the color to the element.

Click OK to generate and view the revised chart in the Chart Viewer. You should now have a custom chart similar to the one shown in figure 48.

Save the chart as a bitmap formatted file. Select File>Save from the menu to specify the location and filename for the file to save, and click Save when done.

Exit the Chart Viewer to return to the Scenarios Manager dialog.

Using the Comparison Tool to generate a report displaying the results for Scenarios 5 through 8.

With these scenarios still selected in the Scenarios list box, click Compare to generate and display the report in the Scenario Comparison Report Viewer dialog as shown in figure 49.

Use the scroll bars and navigational tools located on the toolbar to view different sections of the report as desired.

Exit the viewer to return to the Scenarios Manager dialog. Next, exit the Scenarios Manager dialog and the PM Tool to return to the host application.

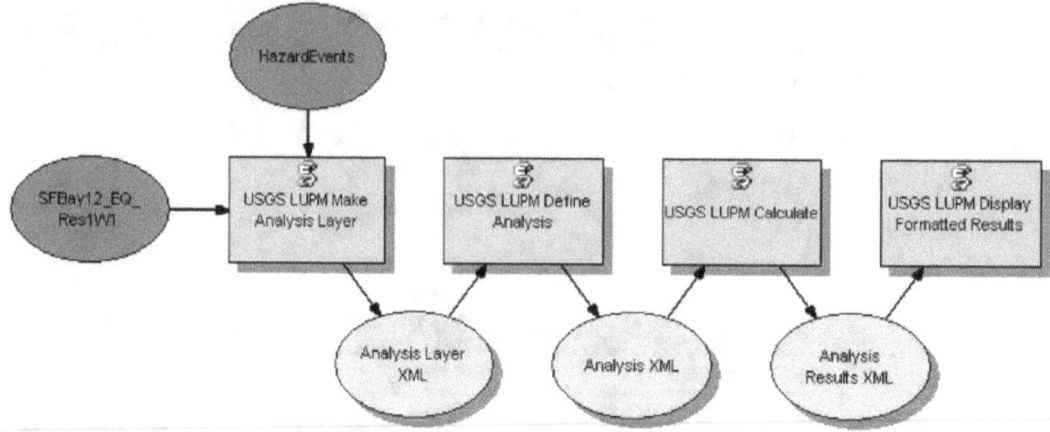

Figure 50. Screen shot of a basic Land Use Portfolio Modeler geoprocessing model.

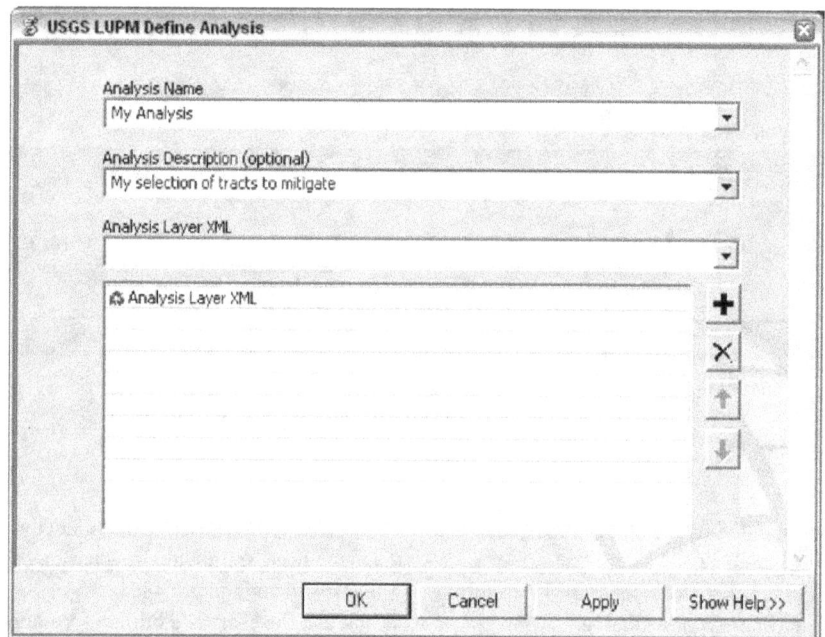

Figure 51.   Screen shot of the in the Define Analysis tool: Editing the analysis name.

LUPM v1.0 Geoprocessing Tools Tutorial

The following tutorial provides examples of how to create and run scenarios using the LUPM v1.0 Geoprocessing Tools package. Some familiarity with the ArcGIS environment is assumed. Prepare for the tutorial as follows:

Launch ArcMap and open the "SFBay12 LUPM example. mxd" file from your copy of the Sample Resources folder.

Make sure that the ArcToolbox tools are available. Select Window>ArcToolbox or click the Show/Hide ArcToolbox button if they are not. Also, make sure that the Favorites tab is selected.

Add the SFBay12 Earthquake example toolbox by right-clicking in the ArcToolbox window and navigating to your copy of the toolbox in your Sample Resources folder.

Part 1: Running the Basic Land Use Portfolio Modeler Geoprocessing Model

Open the SFBay12 Earthquake example toolbox and the Basic Models toolset.

Edit the Basic LUPM Model (fig. 50). Right-click and select Edit (do not select Open as this will open the model for direct execution).

The Basic LUPM Model will show all of the basic LUPM v1.0 geoprocessing components. Run the model by selecting File>Run Entire Model. This action forces all scripts and tools in the model to execute. Note, clicking File>Run or the Run button from the toolbar only executes components that are marked to run, usually those that have

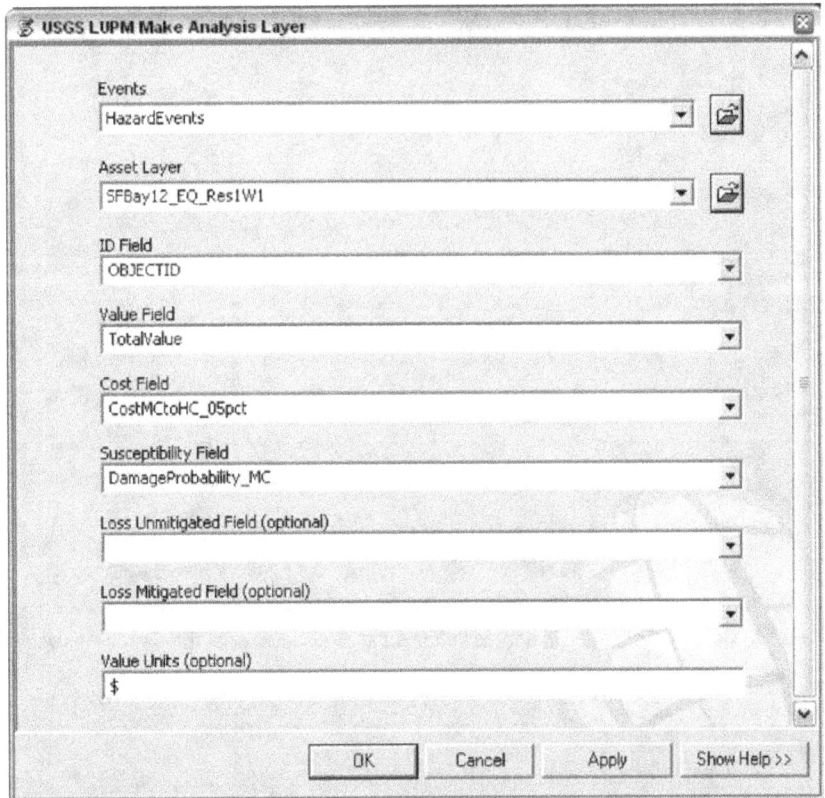

Figure 52. Screen shot showing how to create an analysis layer with the Make Analysis Layer tool.

had their parameters changed or are downstream from a tool that has changed. Tools and data that have been executed are shown with a drop shadow; those that need to be run do not have a drop shadow. Also, this initial scenario does not include any tracts for mitigation investment (a "no mitigation scenario"). The results displayed include the number of locations mitigated (none in this case), and the estimated losses associated with a hypothetical earthquake event.

Return to your ArcMap document and select some tracts for mitigation. Use any of the ArcMap selection methods. The Feature Selection tool is fine for this example, but you can use Select By Attribute or Select By Location as well.

Rerun the entire Basic LUPM Model. You should see locations mitigated and a different estimated loss and estimated wealth.

Note that the description of the analysis is now no longer accurate. Change the name for the analysis by opening the Define Analysis tool (right click the tool on the model diagram and select Open). Type a new name in the Analysis Name text box and update the description in the Analysis Description text box (fig. 51).

Rerun the model, but this time select File>Run, or click the Run button. Only the tools and data sets from Define

Analysis on will be executed, and you should see your new analysis name and description in the results display.

Part 2: Building a Custom LUPM Model

Create a New Geoprocessing Model in the Example Toolbox

Right click the toolbox or toolset and select New>Model. You should see a new, empty model diagram.

Add tools to the model.

Open the USGS LUPM Toolbox (add the toolbox if it is not already available in ArcToolbox's Favorites window; it will be in Toolboxes>System Toolboxes).

Add the Make Analysis Layer, Define Analysis, Calculate, and Display Formatted tools to your model by selecting each tool in the toolbox and dragging it onto your model diagram.

Set the Tool Parameters

Set the parameters for the Make Analysis Layer tool.

Open the Make Analysis Layer tool (fig. 52).

Select the source for the Event ("HazardEvents") from the drop down menu.

Select "SFBay12_EQ_Res1W1" as the Asset Layer.

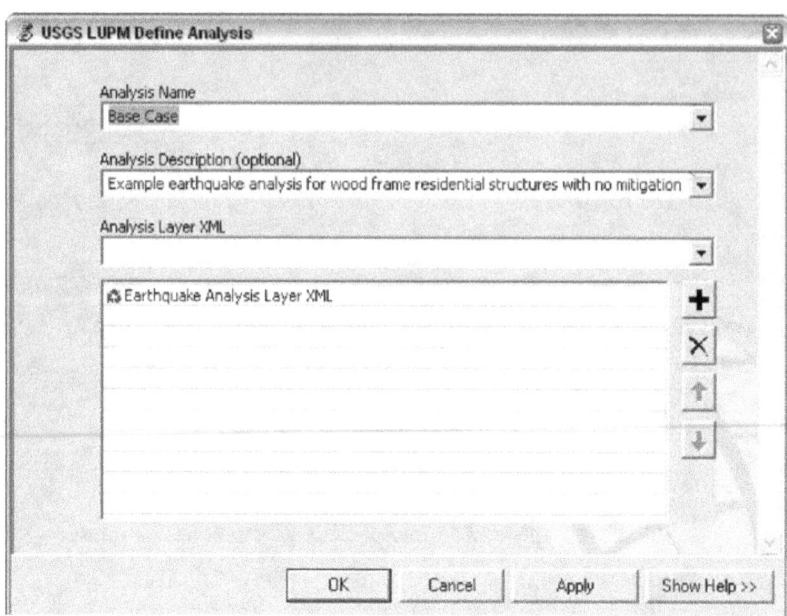

Figure 53. Screen shot showing how to define an analysis layer with the Define Analysis tool.

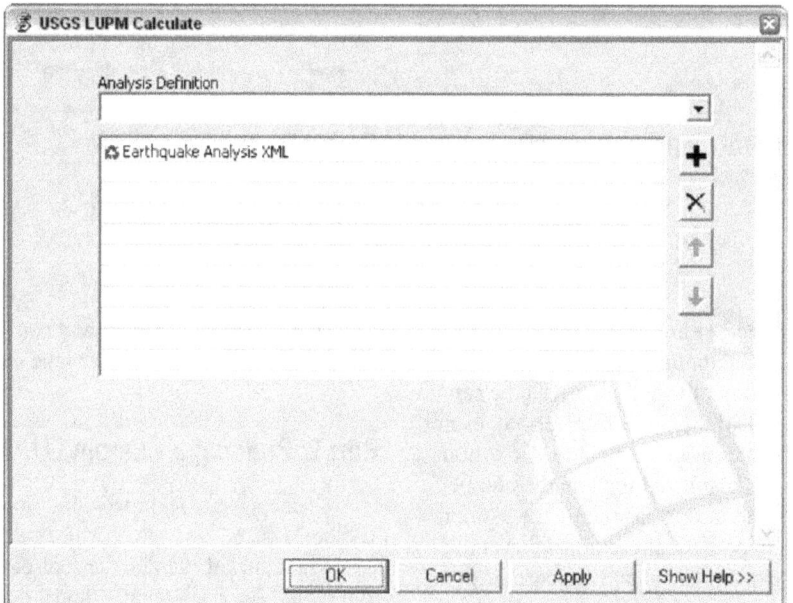

Figure 54. Screen shot showing how to add an analysis to the Calculate tool.

Select "ObjectID" for the ID Field, "TotalValue" for the Value Field, "CostMCtoHC_05pct" as the Cost Field, and "DamageProbabililty_MC" as the Susceptibility Field.

Specify "UnmitigatedLoss_Prop" for the Loss Unmitigated Field and "MitigatedLoss_Prop" for the Loss Mitigated Field.

Click OK to exit.

Rename the output analysis layer XML file by right-clicking the green oval for "Analysis Layer XML" and selecting Rename. Change the name to "Earthquake Analysis Layer XML."

Set the parameters for the Define Analysis tool.

Open the Define Analysis tool (fig. 53).

Specify the analysis name and description.

Select the Analysis Layer XML source from the drop down list. This will be the name of the output from the Make Analysis Layer tool ("Analysis Layer XML" by default; you will have different choices available if you edited the name of the Make Analysis Layer output). Note that you can add more than one analysis layer to the analysis. You can also edit the list of analysis layers (add, delete, reorder). [Note: Be careful if you use the ModelBuilder's Add Connection tool to link the analysis layer with Define Analysis. The analysis layer is

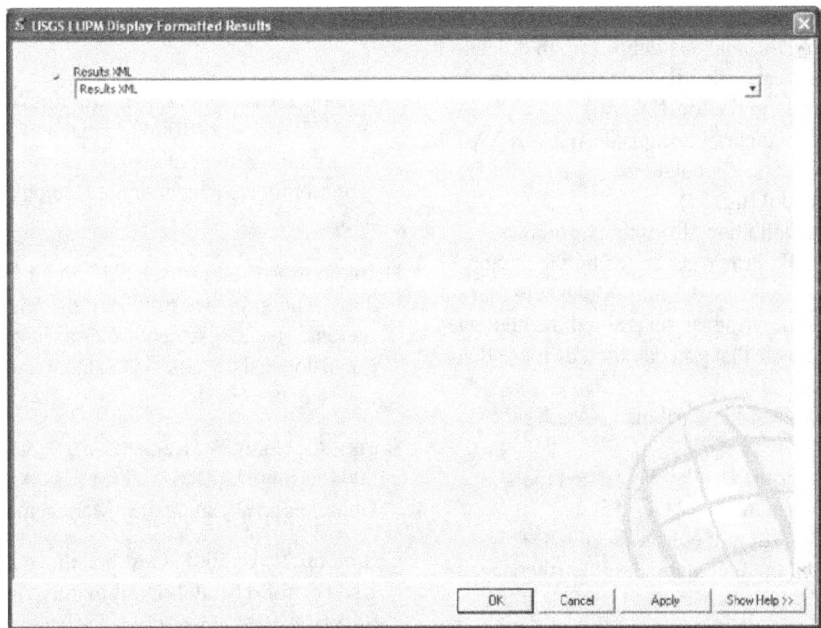

Figure 55. Screen shot showing the selection of the analysis output to display in the Display Formatted tool.

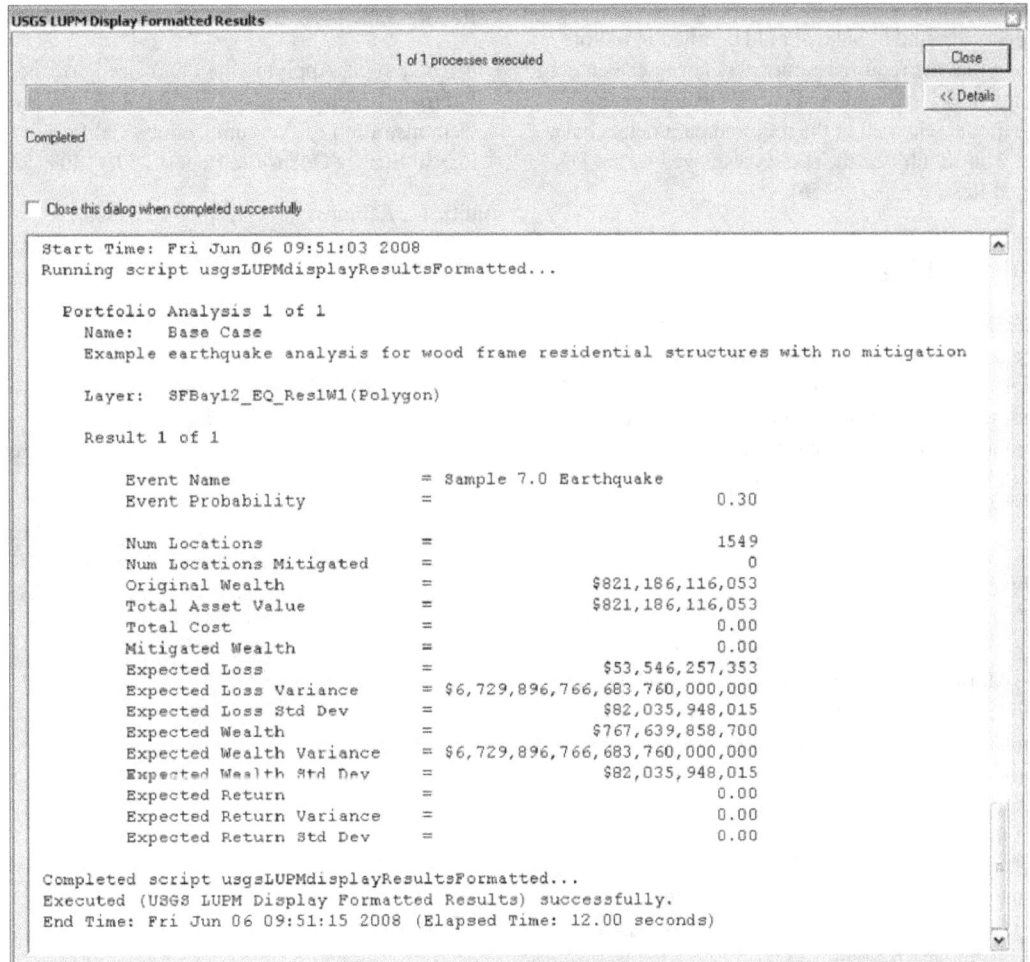

Figure 56. Screen shot of the analysis results as displayed by the Display Formatted Results tool.

an XML string, so, by default, the input will be assigned to the first text parameter in the analysis definition, in this case, the analysis name. Therefore, you will have more control if you add the analysis layers in the tool's form.]

Rename the analysis output "Earthquake Analysis XML."

Set the parameters for the Calculate tool.

Open the Calculate tool (fig. 54).

Select the Analysis Definitions from the drop-down list. These will be the XML strings generated by the Define Analysis tool ("Analysis XML" by default). Make sure that you select the correct XML string, as several will be generated throughout this process. Note that you can include more than one analysis in Calculate.

Rename the results XML to "Earthquake Analysis Results XML."

Set the parameters for the Display Formatted tool.

Open the Display Formatted tool (fig. 55).

Select "Results XML" from the drop-down list. This should be the output from the Calculate tool ("Earthquake Analysis Results XML").

Alternatively, you can use the ModelBuilder's Add Connection tool to associate the results from Calculate with Display Formatted.

Run the Model

All of the boxes and bubbles should be filled in (colored) now, indicating that the required parameters have been entered. If not, open the uncolored tools and check the parameters.

Run the model when all of the required parameters have been entered. You should see the results displayed by the Display Formatted Results tool (fig. 56).

Part 3: Additional Examples

Several additional models are included in the SFBay12 Earthquake example toolbox. Look at these to see more complex examples of LUPM v1.0 geoprocessing models, including multiple analyses and selection of features to mitigate based on attribute and/or location. These models may be copied and modified to add more layers, analyses, and selection criteria.

Acknowledgments

This research was supported by the USGS Geographic Analysis and Monitoring (GAM) program. We thank Murray Journeay and Laura Dinitz for their insightful reviews. LUPM v1.0 is a consolidation of software developed by a number of USGS staff, including contributions from Kyle Gomez, Mark Hessenflow, and Michael Gould. It is also the result of the efforts of the many scientists who have contributed to the land use portfolio methodology, including Richard Bernknopf, Richard Champion, Laura Dinitz, Elizabeth Duffie, Alexander Evans, Paul Hearn, Robb Kapla, Sharyl Rabinovici, David Strong, Anne Wein, and Nathan Wood.

References

Bernknopf, R.L., Dinitz, L.B., Rabinovici, S.J.M., and Evans, A.M., 2001, A portfolio approach to evaluating natural hazard mitigation policies—an application to lateral-spread ground failure: International Geology Review, 43, p. 424-440.

Bernknopf, R., Rabinovici, S.J., Wood, N., and Dinitz, L., 2006, The influence of hazard models on GIS-based regional risk assessments and mitigation policies: International Journal of Risk Assessment and Management, v. 6, nos. 4-6, p. 369-387.

Buriks, C., Bohn, W., Kennett, M., Scola, L., Srdanovic, B., 2004, Using HAZUS-MH for Risk Assessment—How-to-Guide: Federal Emergency Management Agency 433, 117 p.

Champion, R.A., 2005, Cost benefit equations for the Land Use Portfolio Model: Draft manuscript, U.S. Geological Survey Western Geographic Science Center.

Champion, R.A., 2008, A Bernoulli formulation of the Land-Use Portfolio Model: U.S. Geological Survey Open-File Report 2008-1310, 25 p.

Dinitz, L., 2008, Applying the Land Use Portfolio Model to Estimate Natural-Hazard Loss and Risk—A Hypothetical Demonstration for Ventura County, California: U.S. Geological Survey Open-File Report 2008-1309, 12 p.

Dinitz, L., Rabinovici, S., Kapla, R., Taketa, R., Wood, N., and Bernknopf, R., 2003, An interactive GIS linking science to natural hazard mitigation decisions: Proceedings, Urban and Regional Information Systems Association, Annual Meeting, October 11-15, 2003, Atlanta, Georgia, p. 789-799.

Wein, A.M., 2007, Land Use Portfolio Model for natural hazard risk analysis and mitigation decision support: Draft manuscript, U.S. Geological Survey Western Geographic Science Center.

Wein, A.M. and others, 2005, Modeling Natural Hazards in Squamish, British Columbia—A Workshop for planners, emergency preparedness officers and stakeholders in Squamish, September 22, 2005.

Appendix A: Land Use Portfolio Modeler, Version 1.0 Installation

The LUPM v1.0 installer program installs the three packages making up the LUPM v1.0 software: the PM User Control, PM ArcGIS Extension, and LUPM v1.0 Geoprocessing Tools. Each of these packages can be selected for installation during the execution of the installer program. Specific ArcGIS software requirements vary depending on the packages targeted for installation. In addition, the PM ArcGIS extension and LUPM v1.0 Geoprocessing Tools packages require any one of the ArcGIS desktop application suites (ArcInfo, ArcEditor, or ArcView).

PM User Control

Deployment of the PM User Control requires that the ArcGIS Engine Runtime also be installed on the system. This requirement applies to the sample application as well as any application that uses the PM User Control. The LUPM v1.0 software includes a setup program used for installing the ArcGIS Engine Runtime software. Additionally, an ArcGIS product runtime license is required to authorize the use of the ArcGIS Engine Runtime. This requirement can be fulfilled by using an existing license from any installed ArcGIS desktop suite, or by purchasing an ArcGIS Engine Runtime license from ESRI.

PM ArcGIS Extension

The PM ArcGIS Extension is automatically registered to the system at installation. This registration process enables the ArcMap desktop application to recognize the newly installed extension. The extension must also be enabled using the Extensions dialog box, which is available through the Extensions command from the Tools menu in ArcMap. Once the extension is enabled, the toolbar containing the icon to the PM Tool then has to be added to the ArcMap desktop application's toolbar. This task is accomplished using the Customize dialog box, which is available through the Customize command from the Tools menu in ArcMap. Appendix A provides additional instructions regarding the installation of the PM ArcGIS Extension and configuration procedures.

LUPM v1.0 Geoprocessing Tools

Installing the LUPM v1.0 Geoprocessing Tools package will install the USGS LUPM Toolbox into the Toolboxes folder located in the ArcToolBox install directory and the SFBay12 Earthquake Example Toolbox into the Sample Resources folder located in the LUPM Version 1.0 install directory. These toolboxes are automatically loaded with the SFBay12 LUPM example.mxd sample file. They must be added manually to ArcToolBox using the Add Toolbox tool in either ArcCatalog or ArcMap for other map documents.

If added to an ArcMap document, the toolboxes are only available to that document; if added to ArcCatalog, the toolboxes are available to any ArcMap document opened on that system.

Installation Requirements

The following software components need to be installed prior to installing the LUPM v1.0 software:

Microsoft .NET Framework 2.0 (This is available on the ArcGIS 9.2 installation DVD, but is not automatically installed, nor will you find an option to install it. You will need to install it separately if it is not already on your system).

ArcGIS Support Libraries for MS .NET Framework 2.0 (required for PM ArcGIS Extension and LUPM v1.0 Geoprocessing Tools packages).

ArcGIS Desktop v9.2 (required for PM ArcGIS Extension and LUPM v1.0 Geoprocessing Tools packages).

ArcGIS Engine Runtime v9.2 (required for PM User Control package).

Running the LUPM v1.0 Installer Program

Run the setup.exe file located in the LUPM v1.0 Installer folder. Follow the instructions that appear while the setup program is running. Be sure to select the LUPM v1.0 packages that you wish to install. You can also choose to install the sample resources, consisting of the sample data and ArcMap document used for the tutorial.

Post Installation Steps

If you elected to install the PM User Control package during the installation of the LUPM v1.0 software, you will need to check if the ArcGIS Engine Runtime v9.2 is installed on your system. Follow the steps below to check your system for the runtime and install it, if necessary.

Check to see if any other ESRI software products are installed (such as ArcGIS Desktop). If so, check the version numbers of the installed software. The ArcGIS Engine Runtime v9.2 cannot be installed if other ArcGIS software products earlier than v9.2 exist. You will have to remove all ArcGIS software earlier than v9.2 or upgrade the software to v9.2 in order to proceed.

Check if ArcGIS Engine Runtime v9.2 is installed. From the taskbar, select Control Panel>Add or Remove Programs. Look for an entry for the ArcGIS Engine Runtime. If an entry exists, select the entry and click the Support Information control to check that the version number is 9.2.x. If the version is earlier than v9.2, then remove the runtime now.

Run the setup.exe program located in the <LUPM v1.0 install directory>\USGS\LUPM\ArcGIS Engine Runtime folder.

License Requirements

An ArcGIS Engine Runtime v9.2 license is required in order to run applications using the PM User Control. This license is not required, however, if you are a licensed user of other qualifying ESRI software products such as ArcGIS Desktop v9.2. If you are a USGS employee, you may obtain a runtime license by contacting a USGS-ESRI liaison or point-of-contact. If you are not a USGS employee, please contact ESRI directly for further information on purchasing a license.

Appendix B: Sample Dataset Data Layer Attributes

This sample dataset is the basis for the tutorial on using the LUPM v1.0 software. The data set includes census tracts for the 12 counties spanning the San Francisco and Monterey Bay Areas, and is based on the results of an earthquake loss estimation scenario generated using FEMA's HAZUS-MH (version MR2) software. The data attributes described are in the following layers in the dataset: SFBay_12_EQ_Res1W1, AlamedaCo_EQ_Res1W1, and SantaClaraCo_EQ_Res1W1. The following criteria and assumptions were made to create this dataset:

The earthquake scenario is a probabilistic magnitude 7.0 earthquake with a 100-year return.

The mitigation objective for this scenario involved the switch from a moderate building code (MC) to stricter building code (HC).

Census-tract losses for light wood-frame single-family residential structures are derived directly from the HAZUS run.

Damage susceptibility (damage probability) is derived from HAZUS damage proportions (no damage, slight, moderate, extensive, complete). Damage susceptibility = the probability of moderate + extensive + complete loss.

The partial loss values (unmitigated and mitigated) are derived from the HAZUS damage values. Unmitigated losses are the base losses as calculated by HAZUS for single-family residential light wood-frame structures for all building codes. Mitigated losses are those that apply after upgrade of single-family residential light wood-frame structures from the MC to the HC building code. These were calculated by taking the number of structures in the MC class and calculating new losses based on the distribution of the original HC structures across the different damage categories (none, slight, moderate, extensive, complete).

Property value is based on average value per home multiplied by the number of homes in the tract, as derived from the 2000 Census of Population and Housing.

Cost to upgrade structures from MC to HC is estimated at 5 percent and 10 percent of exposure (replacement cost).

The fields in the sample feature class are as follows:

Name	Description
OBJECTID	Object Id internally assigned by ArcGIS.
SHAPE	An ArcGIS field type describing the type of shape the row represents.
Tract	Unique number assigned to the tract.
TotalValue	Total housing value by tract, derived from 2000 Census Average Building value.
Exposure	Potential loss (represented as replacement cost in HAZUS).
CostMCtoHC 05pct	Cost to mitigate (convert from a moderate building code, MC, to a stricter one, HC, at 5 percent of exposure).
CostMCtoHC_10pct	Cost to mitigate (convert from a moderate building code, MC, to a stricter one, HC, at 10 percent of exposure).
DamageProbability MC	Probability of damage given a moderate building code, MC.
UnmitigatedLoss Prop	Proportion of loss for unmitigated tracts. Loss is expressed as a portion of exposure.
MitigatedLoss_Prop	Proportion of loss for mitigated tracts.
LossMC	Tract losses for structures meeting the moderate building code (unmitigated losses).
DamageProbability HC	Probability of damage given a high building code. Applied to upgraded structures.
Loss After Upgrade to HC	Tract losses for MC structures after upgrade to a HC code (mitigated losses). Used to calculate the mitigated loss proportion.

Appendix C: LUPM v1.0 Geoprocessing Tools Documentation

USGS LUPM Calculate

Internal Name:　　usgsLUPMcalculate

Summary:

Performs one or more LUPM analyses.

Multiple analyses may be processed in a single run. However, each analysis is performed independently of other analyses.

The output from this tool is an XML string containing the definition and results for each analysis.

Parameters:

Analysis Definition
Internal name: Analysis_Definition
Direction: Input
Type: Required
Expression: <Analysis_Definition;Analysis_Definition...>
Description: An analysis, as defined by Define Analysis, consisting of one or more analysis layers, as well as an analysis name and description.

Description:

Performs one or more LUPM analyses. The output from this tool is an XML string containing the definition and results for each analysis.

Restrictions:

This tool must be run in conjunction with other USGS LUPM tools.

USGS LUPM Define Analysis

Internal Name:　　usgsLUPMdefineAnalysis

Summary:

Defines an LUPM analysis to be sent to the Calculate tool.

An analysis will consist of one or more analysis layers (XML), as defined by Make Analysis Layer, as well as a name and description.

The output from this tool is an XML string containing the definition for analysis.

Parameters:

Analysis Name
Internal name: Analysis_Name
Direction: Input
Type: Required
Expression: <Analysis_Name>
Description: Name for the analysis. This name will appear in various reports to identify the analysis.

Analysis Description
Internal name: Analysis_Description
Direction: Input
Type: Optional
Expression: {Analysis_Description}
Description: A description of the analysis being performed.

Analysis Layer XML
Internal name: Analysis_Layer_XML
Direction: Input
Type: Required
Expression: <Analysis_Layer_XML;Analysis_Layer_XML...>
Description: An XML string identifying the event, the feature layer, and the fields to be used for an analysis.

Description:

Defines an LUPM analysis. The output from this tool is an XML string containing the definition for analysis.

Restrictions:

This tool must be run in conjunction with other USGS LUPM tools.

USGS LUPM Display Formatted Results

Internal Name:　　usgsLUPMdisplayResultsFormatted

Summary:

Creates a formatted display of the LUPM results generated by the Calculate tool.

Parameters:

Results XML
Internal name: Results_XML
Direction: Input
Type: Required

Expression: <Results_XML>
Description: An XML string containing the results from the LUPM calculations. This will be the output from the Calculate script.

Description:

Creates a formatted display of the LUPM results generated by the Calculate tool.

Restrictions:

This tool must be run in conjunction with other USGS LUPM tools.

USGS LUPM Get Feature Class

Internal Name: usgsLUPMgetFeatureClass

Summary:

Returns the feature class associated with a feature layer in an ArcGIS application.

The output is a string representing the path to the feature class.

This tool is provided to resolve an issue in creating new feature layers from existing feature layers. The Make Feature Layer tool returns a reference to the existing feature layer, not a new feature layer. As a result, changes to the newly-created feature layer also change the selections in the original feature layer.

Parameters:

Feature Layer
Internal name: Feature_Layer
Direction: Input
Type: Required
Expression: <Feature_Layer>
Description: A feature layer in an ArcGIS application.

Description:

Returns the feature class associated with a feature layer in an ArcGIS application. The output is a string representing the path to the feature class.

Restrictions:

None.

USGS LUPM Make Analysis Layer

Internal Name: usgsLUPMmakeAnalysisLayer

Summary:

Creates an LUPM analysis layer as an XML string.

The event represents a specific natural environmental hazard, such as an earthquake or flood. The feature layer represents the assets to be modeled and the fields represent such data as the value of the feature, the cost to mitigate the event, the susceptibility of the feature to damage from the event.

Generally, the events associated with a analysis layer should represent those for which the selected fields are appropriate. For example, the susceptibility to damage and the cost to mitigate will be associated with a specific event, such as a magnitude 7.1 earthquake. Therefore, the values in the selected fields should be appropriate for the selected event(s). Optional fields include values for loss if mitigated and loss if not mitigated. These provide the opportunity to specify partial losses.

The output from this tool is an XML string containing the specifications for the analysis of the feature layer. The analysis layer XML is used in subsequent LUPM calculation operations.

Parameters:

Event
Internal name: Event
Direction: Input
Type: Required
Expression: <Event>
Description: Event (e.g., earthquake, flood, etc.) triggering the impacts to be mitigated. The event data includes the name for the event (Name field) and the event probability (Probability field). A description of the event (Description field) is optional.

Asset Layer
Internal name: Asset_Layer
Direction: Input
Type: Required
Expression: <Asset_Layer>
Description: Feature layer containing the asset information. Fields containing values for the analysis may be part of this layer, or may be joined to it.

ID Field
Internal name: ID_Field
Direction: Input
Type: Required
Expression: <ID_Field>
Description: Identifier used to track specific features. Should be unique for each feature.

Value Field
Internal name: Value_Field
Direction: Input
Type: Required

Expression: <Value_Field>
Description: Value of the asset (feature).

Cost Field

Internal name: Cost_Field
Direction: Input
Type: Required
Expression: <Cost_Field>
Description: Cost to mitigate the effects of the specified event.

Susceptibility Field

Internal name: Susceptibility_Field
Direction: Input
Type: Required
Expression: <Susceptibility_Field>
Description: The susceptibility of loss from the specified event, expressed as the probability of loss. Must be a value between 0 and 1.

Loss Unmitigated Field

Internal name: Loss_Unmitigated_Field
Direction: Input
Type: Optional
Expression: {Loss_Unmitigated_Field}
Description: Optional value indicating a partial loss for a feature if the feature is not selected for mitigation and suffers damage. Expressed as a proportion of value. If this field is missing, then the analysis assumes complete loss if the feature is not selected for mitigation and is impacted by the event.

Loss Mitigated Field

Internal name: Loss_Mitigated_Field
Direction: Input
Type: Optional
Expression: {Loss_Mitigated_Field}
Description: Optional value indicating the loss for a feature if the feature is selected for mitigation and suffers some damage. Expressed as a proportion of value. If this field is missing, then the analysis assumes no loss if the feature is selected for mitigation.

Value Units

Internal name: Value_Units
Direction: Input
Type: Optional
Expression: {Value_Units}
Description: Optional units for value and cost (e.g., U.S. $). Incorporated into the results XML and is used for reporting.

Description:

Creates an LUPM analysis layer from a feature layer, an event (e.g., an earthquake), and a set of fields in the feature layer associated with the event. The output from this tool is an XML string containing the specifications for the analysis of the feature layer.

Restrictions:

This tool must be run in conjunction with other USGS LUPM tools. In particular, this tool uses the "getFeatureLayer-AsXML_Function" geoprocessing function tool.

USGS LUPM Write Results

Internal Name: usgsLUPMwriteResults

Summary:

Write the results from the Calculate script to a text file.

Parameters:

Analysis Results

Internal name: Analysis_Results
Direction: Input
Type: Required
Expression: <Analysis_Results>
Description: An XML string containing the results from the LUPM calculations. This will be the output from the Calculate script.

Result File

Internal name: Result_File
Direction: Input
Type: Required
Expression: <Result_File>
Description: Text file to which the results will be appended. The file must exist.

Description:

Write the results from the Calculate script to a text file.

Restrictions:

This tool must be run in conjunction with other USGS LUPM tools.

USGS LUPM Write XML

Internal Name: usgsLUPMwriteXML

Summary:

Write the results from the Calculate script to an XML file.

Parameters:

Analysis Results

Internal name: Analysis_Results
Direction: Input
Type: Required

Expression: <Analysis_Results>
Description: An XML string containing the results from the LUPM calculations. This will be the output from the Calculate script.

XML File
Internal name: XML_File
Direction: Output
Type: Required
Expression: <XML_File>
Description: The output XML file.

Description:

Write the results from the Calculate script to an XML file.

Restrictions:

This tool must be run in conjunction with other USGS LUPM tools.

USGS Function Get Feature Layer as XML

Internal Name: usgsFunctionGetFeatureLayerAsXML

Summary:

Returns a feature layer as an XML string.

Retains joins and feature selections defined for the feature layer.

The tool is a geoprocessing function tool designed to be called from Python script.

Parameters:

Feature Layer
Internal name: feature_layer
Direction: Input
Type: Required
Expression: <Feature Layer>
Description: Feature layer as part of an ArcMap document or created in a model or script.

Feature Layer XML
Internal name: feature_layer_xml
Direction: Output
Type: Optional
Expression: {LUPM_Results_XML}
Description: Feature layer converted to an XML string. This parameter is declared 'optional' to satisfy the Geoprocessor user interface. The results will be inserted here automatically.

Description:

Returns a feature layer as an XML string. Retains joins and feature selections defined for the feature layer.

Restrictions:

This tool must be run in conjunction with other USGS LUPM tools.

USGS Function LUPM Calculate

Internal Name: usgsFunctionLUPMCalculate

Summary:

Performs the portfolio calculation.

The calculation is based on the analysis created in Define Analysis.

The tool is a geoprocessing function tool designed to be called from Python script.

Parameters:

LUPM Analysis XML
Internal name: LUPM_Analysis_XML
Direction: Input
Type: Required
Expression: <Analysis;Analysis...>
Description: An XML string containing the analysis definition, as created by Define Analysis.

LUPM Results XML
Internal name: LUPM_Results_XML
Direction: Output
Type: Optional
Expression: {LUPM_Results_XML}
Description: LUPM results incorporated into an XML string. This parameter is declared 'optional' to satisfy the Geoprocessor user interface. The results will be inserted here automatically.

Description:

Performs the portfolio calculation. The calculation is based on the analysis or analyses created in Define Analysis.

Restrictions:

This tool must be run in conjunction with other USGS LUPM tools.

Appendix D: LUPM v1.0 Core Libraries

Name	Description
PM_BLL	Contains the implementation of the business logic layer, which serves as an intermediary between a given graphical user interface and the data access layer. It enables transactions against a PM Database to be performed transparently. This layer is comprised of operations serving each of the data structures defined in the PM_Common library. Each data structure includes a set of create, retrieve, update, and delete (CRUD) operations, which are performed against an associated table in a PM Database.
PM_Chart	Contains the implementation of the Charts tool included with the PM Tool. This tool produces one of two charts using data derived from one or more scenarios. One chart provides a comparison of up to two output variables (such as expected losses or expected avoided losses) along with their standard deviations for one or more scenarios. The other chart provides a comparison of ancillary benefit outputs for one or more scenarios.
PM_Common	Contains the definition of various data structures, which hold data passed to and from an underlying PM Database and are used throughout the core set of libraries. A brief description of each of these data structures follows.
	ABParameter contains data associated with an ancillary benefit input parameter used in a scenario.
	ABResult stores output associated with an ancillary benefit generated from a scenario.
	HazardClass holds data describing a hazard class.
	HazardEvent holds data describing a hazard event.
	PMFeature stores scenario output pertaining to a feature.
	PMParameters stores input parameters used in a scenario.
	PMResults stores the output for a scenario, formatted for report purposes.
	Scenario contains data describing a scenario.
	ScenarioLayer contains data describing a layer used in a scenario.
	ScenarioParameters contains the parameters to be used in a scenario.
	ScenarioResults stores the output returned from a scenario.
	ScenarioSet contains data describing a scenario set.
PM_Controls	Contains the implementation of miscellaneous user controls used in various interfaces included in the core set of libraries.
PM_DAL	Contains the implementation of the data access layer, which enables data to flow through between the business logic layer and a given PM Database. The layer includes the data access protocols and structured query language (SQL) commands required to create and access a PM Database, as well as to perform transactions against it.
PM_Engine	Contains the implementation of menu and toolbar items included in the PM User Control, as well as the control itself.
PM_ESRI	Provides a number of utility functions for manipulating or processing ESRI-based objects such as retrieving feature layer data from a given ArcMap document source.
PM_Math	Contains the logic used to implement the LUPM, providing functions that prepare the LUPM input data, perform the LUPM calculations, and return results in various formats.
PM_Reports	Implements the Reports tool included with the PM Tool. This tool produces a Crystal Reports-based report, which compares the outputs of one or more scenarios. Includes the Crystal Reports definition file for this report.
PM_UI	Contains the graphical user interfaces used in the PM Tool implementation of the LUPM.

Appendix E: PM Database Table Objects

Table—ABParameters
[Contains rows representing ancillary parameter inputs used in scenarios. Each row represents an ancillary benefit parameter used as input for a particular scenario. Row must be associated with a specific row in the ScenarioParameters table.]

Name	Type	Required	Description
Id	Number	Yes	Unique identifier value assigned to this row.
Name	String	Yes	Name of ancillary benefit (cultural, critical, economic, environmental, or safety).
Layer	String	Yes	Layer containing this ancillary benefit field.
ParcelIdField	String	Yes	Parcel Id or another primary identifier field for layer.
Field	String	Yes	Field containing the ancillary benefit.
ScenarioParameterId	Number	Yes	Foreign key value to the ScenarioParameters table.

Table—ABResults
[Contains rows representing scenario ancillary benefit results. Each row contains results of a specific ancillary benefit for a particular scenario and must be associated with a specific row in the ScenarioResults table.]

Name	Type	Required	Description
Id	Number	Yes	Unique identifier value assigned to this row.
Name	String	Yes	Name of ancillary benefit (cultural, critical, economic, environmental, or safety).
NumLocations	String	Yes	Number of locations identified as having this benefit.
NumLocationsMitigated	String	Yes	Number of mitigated locations identified as having this benefit.
ScenarioResultId	Number	Yes	Foreign key value to the ScenarioResults table.

Table—HazardClasses
[Contains rows representing hazard classifications or types.]

Name	Type	Required	Description
Id	Number	Yes	Unique identifier value assigned to this row.
Name	String	Yes	Name of hazard class.
Description	String	No	Description of hazard class.

Table—HazardEvents
[Contains rows representing specific hazard events. Each row pertains to a hazard event, which is associated with a specific hazard class.]

Name	Type	Required	Description
Id	Number	Yes	Unique identifier value assigned to this row.
Name	String	Yes	Name of hazard event.
Description	String	No	Description of hazard event, which, perhaps, may include describing the severity and frequency of event, as well as the geographical area affected.
Probability	Number	Yes	Probability of event occurring expressed as a value between 0 and 1.
HazardClassId	Number	Yes	Foreign key value to the HazardClasses table.

Table—ScenarioLayers
[Contains rows representing specific hazard events. Each row pertains to a hazard event, which is associated with a specific hazard class.]

Name	Type	Required	Description
Id	Number	Yes	Unique identifier value assigned to this row.
Name	String	Yes	Name of a layer used in a specific scenario.
SelectedIds	String	No	A concatenated string containing the Ids of selected features in the layer.
ScenarioId	Number	Yes	Foreign key value to the Scenarios table.

Table—ScenarioParameters
[Contains rows representing parameter inputs for scenarios. Each row identifies parameter inputs used for a particular scenario.]

Name	Type	Required	Description
Id	Number	Yes	Unique identifier value assigned to this row.
EventProbability	Number	Yes	Event probability expressed as a value between 0 and 1, inclusive. This value could differ from that originally assigned to the event.
PrimaryParcelLayer	String	No	Layer containing most, if not all, fields, which can be used as inputs to the LUPM.
PrimaryParcelLayerParcelIdField	String	No	Parcel Id or another identifier field for Primary Parcel Layer.
AssetLayer	String	Yes	Layer containing asset value field.
AssetParcelIdField	String	Yes	Parcel Id or another primary identifier field for asset layer.
AssetField	String	Yes	Asset value field.
SusceptibilityLayer	String	Yes	Layer containing damage susceptibility field.
SusceptibilityParcelIdField	String	Yes	Parcel Id or another primary identifier field for susceptibility layer.
SusceptibilityField	String	Yes	Damage susceptibility field.
CostLayer	String	Yes	Layer containing mitigation cost field.
CostParcelIdField	String	Yes	Parcel Id or another primary identifier field for cost layer.
CostField	String	Yes	Mitigation cost field.
PcntInitLossLayer	String	No	Layer containing the extent of loss without mitigation (percent initial loss) field.
PcntInitLossParcelIdField	String	No	Parcel Id or another primary identifier field for the extent of loss without mitigation layer.
PcntInitLossField	String	No	Extent of loss without mitigation field. Valid values for this field range between 0 and 1, inclusive.
PcntMitLossLayer	String	No	Layer containing the extent of loss (percent mitigated loss) field.
PcntMitLossParcelIdField	String	No	Parcel Id or another primary identifier field for the extent of loss with mitigation layer.
PcntMitLossField	String	No	Extent of loss with mitigation field. Valid values for this field range between 0 and 1, inclusive.
OtherMitCosts	Number	No	Mitigation costs applied regionally (community mitigation costs).
HazardEventId	Number	No	Foreign key value to the HazardEvents table.
ScenarioId	Number	Yes	Foreign key value to the Scenarios table.

Table—Scenarios

[Contains rows representing scenarios, which were created, run, and saved via the PM Tool. A scenario must be associated with a specific scenario set.]

Name	Type	Required	Description
Id	Number	Yes	Unique identifier value assigned to this row.
Name	String	Yes	Name of scenario.
Description	String	No	Description of scenario, which, perhaps, may describe the hazard event, mitigation strategy, input parameters, data assumptions, and other relevant information affecting scenario outcome.
CreateDate	Date/Time	Yes	Date on which scenario record was created.
ScenarioSetId	Number	Yes	Foreign key value to the ScenarioSets table.

Table—ScenarioSets

[Contains rows representing scenario sets. A scenario set is used as a way of organizing scenarios. Preferably, a scenario set should be created and used to associate with scenarios for the same hazard event.]

Name	Type	Required	Description
Id	Number	Yes	Unique identifier value assigned to this row.
Name	String	Yes	Name of scenario set.
Description	String	No	Description of scenario set, which, perhaps, may describe the scenarios it will be associated with.
CreateDate	Date/Time	Yes	Date on which scenario set record was created.

www.ingramcontent.com/pod-product-compliance
Lightning Source LLC
Chambersburg PA
CBHW082031190526
45166CB00017B/2964